T0221150

Native Soil

NATIVE SOIL

A HISTORY OF THE
DEKALB COUNTY FARM BUREAU

Eric W. Mogren

NORTHERN ILLINOIS UNIVERSITY PRESS / DEKALB

© 2005 by Northern Illinois University Press

Published by the Northern Illinois University Press, DeKalb, Illinois 60115

Design by Julia Fauci

Library of Congress Cataloging-in-Publication Data

Mogren, Eric W. (Eric William)

Native soil : a history of the DeKalb County Farm Bureau / Eric W. Mogren.—

   p.   cm.

Includes bibliographical references and index.

ISBN-13: 978-0-87580-348-7 (clothbound: alk. paper)

ISBN-10: 0-87580-348-2 (clothbound: alk. paper)

1. DeKalb County Farm Bureau (DeKalb County, Ill.)—History. I. Title.

HD1485.D45M64 2006

338.1'0974365—dc22

2005009598

For My Father

CONTENTS

ACKNOWLEDGMENTS

• When I arrived in DeKalb County, Illinois, ten years ago, I felt as if I was a long way from home. I spent the first thirty years of my life in northern Colorado, where the high plains meet the mountains. The soil there, at the base of the Rockies, is tan-colored and thin. Husbandry in that semiarid region focused on raising winter wheat, livestock, or-chards, and, when I was a boy, sugar beets. Family-owned ranches com-peted with industrial feed lots. Truck farms grew vegetables for regional grocery stores. Hardscrabble homesteads dotted the foothills and moun-tain valleys. Farming was possible only because an elaborate network of hundred-year-old irrigation canals and feeder ditches that snaked through the countryside delivered water from storage reservoirs to lucky fields—winter mountain snow, not spring rain, determined the success or failure of autumn harvests. Deep wells and center-pivot sprinklers cre-ated huge perfect circles of green on the brown landscape. Windmills pumped water into steel stock troughs to quench the thirst of sun-baked cattle and sheep.

Like many Americans, some of my ancestors were farmers. I was raised with stories about my mother's Icelandic family, who pioneered in southern Manitoba and southwestern Minnesota. Growing up on the high plains and among the towering mountains of the American West, however, meant that those tales of joy and sadness, birth and death, hardship and success, took place in a midwestern landscape as foreign to me as the subarctic Icelandic farmsteads and volcanic peaks my ances-tors left behind forever in the late 1870s. When I was a boy, we made a few trips from Colorado to the family farm in Minnesota, named "Grund," a shortened version of the name of the original Icelandic farmstead "Grundarholl," meaning "farm on a hill." My memories of Grund, however, are a kaleidoscope of such childish impressions as the pale blue color of flax fields, the thrill of riding on a tractor, the farm an-imals, the mystery of the woodlot, the back staircase and attic in the main house, and the smell of the kitchen. The old men and women of Grund I met as a child, who still spoke Icelandic to each other from time to time, included my grandmother and great-uncles. Today they are long gone, and photographs of stern men and women dressed in dark suits and dresses are the only links we have to those pioneers. They were the first native-born Americans on my maternal side, and their lives coincided with the greatest and most turbulent period in the history of American agriculture. As I discovered, their lives were not so very different from the lives of farmers of DeKalb County, Illinois.

Still, despite my early introduction to the Midwest, I was overwhelmed by the flatness of the DeKalb County countryside, the wide-open sky, the linear horizon, the garden-like fields, and the tyranny of the annual agricultural cycle. I experienced emotions that have been common to three centuries of newcomers to the DeKalb area—the solitude of open space coupled with amazement over the breathtaking productivity of the deep, black soil. I was an alien lost in a strange place. I discovered that this historical project was as much a journey of personal exploration of my new home as a presentation of the history of one of the nation's most successful Farm Bureaus. Although DeKalb County is not my native soil, I have come to love my adopted land.

This is a story about farsighted men and women in one county deep in the heart of America's corn belt. It could not have been completed without generous assistance from institutions, colleagues, old and new friends, and family. Several of them deserve special recognition.

I am indebted to the DeKalb County Farm Bureau for making this book possible. The Farm Bureau Foundation welcomed the project and provided financial assistance. It was a delight to work with everyone at the Bureau, but especially Doug Dashner, the Bureau's manager, who opened the Bureau's historical materials to me and gave me a free hand to sort through it all. Sherry Johnson and Mariam Wassmann at the Bureau assisted with the illustrations and helped me locate important historical information. I also want to thank everyone in the Farm Bureau "family" who shared their personal Bureau memories with me. Their enthusiasm for this project is exceeded only by the kindness they showed me.

My Northern Illinois University colleague David Kyvig, a strong advocate for local history, persuaded me to take the project and offered sound advice and reassurance along the way. Other special friends in the history department helped me keep a good sense of humor while balancing my research and academic responsibilities. Thanks also to the university for awarding my sabbatical leave that enabled me to complete my research and an early draft of the book. The staff at the Regional History Center and at Founders' Memorial Library were very helpful, as usual. Mary Lincoln, at the Northern Illinois University Press, was an early supporter of the project and a remarkably patient editor. Several referees offered insightful critical comments and recommendations that improved the final version of the book. Mary and her Press colleagues did a wonderful job of steering the book through all the publication hurdles.

I also wish to thank the Charles Roberts family. Charlie's father-in-law, Tom Roberts, Sr., was one of the early farm advisors, and for three generations the Roberts family has been closely associated with the Bureau. Charlie knew that the history of the DeKalb County Farm Bureau was important, wanted it to be told, and convinced me and the NIU Press to undertake the book.

Elroy "Al" Golden deserves special recognition. Al became the DeKalb County Extension Advisor in May 1954 and, until his resignation sixteen years later, was an important advisor to the Bureau's leadership during some of the most exciting and turbulent times in the organization's history. Al also recognized the importance of preserving the Bureau's history and for years collected and preserved historical materials relating to the Bureau. Those materials are now the core of the Bureau's historical archive. Several years ago Al also produced a report of the Bureau's history, which proved to be enormously helpful to me in this project. He was generous with his knowledge and gracious with his time. It is not an exaggeration to say that the history of the DeKalb County Farm Bureau would have been lost had not Al persevered in his labor to save the remaining fragments of the DeKalb County Farm Bureau's historical material.

I owe the most to my family. My mother taught me the delights of scholarship and personal satisfaction that comes from being a "lifelong learner." I regret that my father did not live long enough to see this book completed. I think he would have enjoyed it. My father-in-law, Arnold Gurwitz, is a gentle and unselfish man who has always stood by me. DeKalb County is the only home my children, Leif and Claire, remember. They are both delightful and constantly remind me about the things in life that are truly important. Above all, I want to thank my beloved wife, Linda, who has been at my side through twenty wonderful years.

NATIVE SOIL

INTRODUCTION

- DeKalb County, Illinois, is located in north-central Illinois, about twenty-five miles south of the Wisconsin border and about fifty miles west of the shores of Lake Michigan. Shaped like a rectangle thirty-six miles north to south and eighteen miles east to west, the county covers roughly 407,000 acres, about 633 square miles. Like many places in America's midwestern heartland, the flat landscape has few natural geological features that can be used to define conveniently political borders between counties. Consequently, its boundaries run in the perfectly straight lines and square angles first defined by early government surveyors. The county is subdivided into nineteen townships, each with equally plumb boundaries.[1]

The rigid, arbitrary, and recent subdivisions that scribe DeKalb County's surface serve our human desire to define and control land, but have nothing to do with the region's complex underlying geology. The pre-Cambrian granite bedrock lying about 4,000 feet below DeKalb County's rich soil is the foundation of the county and is some of the oldest rock on the planet. Above that, prehistoric oceans deposited layer upon layer of sandstone, shale, and limestone to a thickness of around 2,500 feet. These layers provide an aquifer for the deepest wells in the county, and fossil hunters easily find the remains of prehistoric plants and marine animals in the limestone quarries that dot the region. Four major periods of glaciation also left marks on the land. Beginning in the Pleistocene era, about 55 million years ago, the earth's temperature gradually cooled. Snow and ice accumulated over millions of years to form continent-size sheets of flowing ice that reached as far south as today's Ohio and Missouri River valleys. The slowly creeping glaciers repeatedly scoured the planet's surface and moved countless millions of tons of pulverized rock great distances. The motion and incalculable weight of the glaciers sheared off geological features and left behind broad valleys, marshes, and lake bottoms as they melted away. The Great Lakes are the most obvious glacial ghosts on the continent. Each time the ice stopped moving, the debris plowed up by the ice sheets and the unconsolidated material dropped by the melting glaciers and washed out by the melt waters created assorted layers and textures of subsurface strata. Such glacial material varies in depth throughout DeKalb County, from 40 to 450 feet.

Millions of years of repeated glacial action and interglacial erosion also sculpted DeKalb County's gentle topography. The debris the glaciers pushed ahead of themselves formed three broad, low rises, called moraines, that run roughly southwest to northeast across the county. Even today, the subtle, undulating ridges of the Bloomington and Cropsey moraines are identifiable to observant travelers driving east to west across the county. The Farm Ridge moraine, in the southeast part of the county, is the most identifiable geographical feature in the area, although it is more difficult to distinguish from the surrounding countryside than the county's other two moraines. The highest elevation in the county is 977 feet above sea level, near the hamlet of Lee at the extreme western boundary of the county; the lowest outside of Sandwich in the county's southeast corner is 645 feet. There are a few gentle hills, with the greatest topographical change occurring in Paw Paw Township, where there is an altitude change of about 180 feet in five and a half miles. Generally, however, the county is flat—nearly 96 percent of the county has a slope of 7 percent or less. Most of the northern part of the county is drained by the Kishwaukee River, a small stream most of the time, which rises near Shabbona and meanders northwest until it drains into the Rock River. The southern part of the county is in the Fox River watershed.[2]

DeKalb County is located deep in the interior of North America. Consequently, its continental, temperate climate swings seasonally between warm summers and cold winters. Average temperatures are in the seventies Fahrenheit during the summers and twenties during the winter months, but extreme temperatures are not uncommon. The highest recorded temperature was 109 degrees, the lowest 28 degrees below zero; the greatest difference in seasonal high and low temperatures occurred in 1933, when the year's recorded high was 100 degrees and the low was minus 26. Killing frosts usually disappear in early May and return again in early October, providing area farmers with a generous growing season of about 160 days. Annual precipitation has topped fifty inches, but the average is about thirty-five inches, with most falling as rain between

May and September. Thunderstorms occur about thirty or forty times a year, mostly during the summer. Snowfalls seldom exceed twelve inches. For most of the year, the wind blows out of the west or south, pushing the high clouds that partly or completely shade the county nearly three-quarters of the time.[3]

DeKalb County's most important natural resource is its soil, which owes its existence and fertility to eons of geological and climatic cycles. Each of the major glaciers that periodically covered the county left behind thick deposits of boulders, gravel, sand, silt, and clay. During the interglacial periods, the prevailing west winds carried the particles of rock dust, produced by the grinding action of the glaciers, from the flood plains of the ancient Mississippi River, and other areas even further west, and deposited it over the substrata of rock and gravel to a depth averaging about eighteen inches throughout the county. This windblown material, called "loess," resulted in the fairly uniform and stable brown, black, and yellow silt loam soils that cover more than 90 percent of DeKalb County. These loam soils, which are composed predominantly of particles that are finer than sand and larger than clay, are water permeable and rich in chemical nutrients that plants require for healthy growth.

The soil characteristics and chemistry supported a lush cover of prairie grasses and deciduous forests during the twelve thousand years since the end of the last ice age that further influenced soil development. The open plains of DeKalb County was home to a large number of plant species, including Indian and switch grasses, and the predominant plants of the climax prairie ecosystem, big bluestem grass *(Andropogon gerardi)* and little bluestem grass *(Andropogon scararius)*. Big bluestem grass was especially impressive, with a root system that penetrated six or seven feet into the soil and that grew, by some early reports, as tall as a man on horseback. More than 90 percent of DeKalb County was once covered by these luxuriant prairie grasses.[4] Bottomland and upland forests of cottonwood, soft maple, elm, hickory, and white, red, and burr oak covered only about 8 percent of the county. The constant cycle of growth and decay of prairie plants contributed organic matter that created a deep layer of topsoil. Microscopic flora and fauna within the soils produced complex organic compounds that encouraged a diverse prairie ecosystem and further enriched the soil. Frequent fires sparked by lightning, and later accidentally and deliberately by humans, burned off the plants and layers of humus at the upper layers of the soil, releasing nutrients and clearing the way for renewed plant growth. Under the prevailing climatic conditions in DeKalb County, the prairie ecosystem built topsoil at the rate of one inch every 100 to 400 years, resulting in a layer of rich, garden-like topsoil several feet thick over most of the area.[5]

The prairie also supported a diverse population of animals. When white settlers first arrived in DeKalb County, bison, elk, and white-tailed deer grazed on the open prairies and along the edges of the timbered lands. Raccoon, river otter, beaver, opossum, weasel, badger, woodchuck, mink, skunk, squirrel, and muskrat were the most common small furbearers. The waterways and low marshes were breeding grounds and migratory resting places for geese, cranes, snipes, rails, and mallard, blue wing teal, wood, and ruddy ducks. Upland birds, such as the woodcock, mourning dove, ruffed grouse, prairie chicken, wild turkey, and bobwhite quail were common sights as late as the middle nineteenth century.

For thousands of years, the DeKalb area was home, too, to prehistoric Amerindian hunters who probably arrived in the area some 11,000 to 12,000 years ago. Between 2,500 and 3,500 years ago, far to the south in what is now Mexico, prehistoric people first domesticated maize, or "corn," from a weedy grass, called "teosinte."[6] Over the centuries, knowledge of its cultivation gradually spread northward, probably following the major north American river systems, including the Illinois River, and Native American overland trade routes into the temperate prairie zones. Some of the earliest evidence of corn cultivation in Illinois dates to about 100 CE from the Hopewellian sites along the Illinois and Mississippi rivers.[7] By 1000 CE, the regular cultivation of corn, beans, squash, and melons by the prehistoric Mississippian culture peoples in the region provided them with the agricultural security that resulted in homesteads, camps, and villages near timber and water sources. These peoples also probably supplemented their diets by hunting. At some locations, such as Cahokia, the domestication of corn and other plants resulted in a complex urban center whose cultural networks spread outward to influence even the most remote villages on the fringes of their territories. In DeKalb County, early white settlers found evidence of prehistoric human habitation, including stone axes, pestle stones for grinding corn and other hard-shelled foods, knives, pipes, and projectile points. In the past, such items turned up regularly in freshly plowed furrows around the county, and many sharp-eyed farmers amassed impressive artifact collections. Although less common today, it is still not unusual to find such objects in the area.

During the sixteenth and seventeenth centuries, European explorers and settlers brought new influences to the northern Illinois region that dramatically altered the farming and hunting lifestyles of the Native Americans living there. Horses, European trade items, and Old World diseases devastated traditional Native American cultures. The fur trade, in particular, contributed to intertribal tension, competition, and even warfare, as tribes sought to defend their hunting territories and acquire those of their neighbors. At the time of white contact in the Illinois region in the late 1690s, the Algonquian-speaking Asa-ki-waki (People of the Outlet) and Mesquakie (Red Earth People) tribes, or as the French called them, the "Sauk" and "Fox" tribes, farmed and hunted the prairies and timbered streambeds of northern Illinois. Their principal crop was corn, although they also raised squash and beans; bison and smaller game provided meat and other animal products. During the eighteenth century, the Pottawattomi, another Algonquian-speaking tribe, gradually moved into the territory south of Lake Michigan and expanded westward until, by the early 1800, they were the predominate Native American population in DeKalb County.

The first Europeans to reach Illinois were the French, whose earliest reports of the region describe it as a paradise of grass. For the hearty French explorers, *couriers de bois,* and Jesuit missionaries, all of whom were most at home in their cold northern European and Canadian forests, the promise of the new region was overwhelming. Game and wildfowl were abundant. "Today we saw over 50 bears," one Jesuit wrote, "and of all that we killed we took only 4, in order to obtain some fat." Jacques Marquette commented extensively on the fruitful beauty of the area; Zenobius Membre noted the "richness and fertility of the plains" that provided the local peoples with crops. Robert Cavelier de La Salle stressed the future possibilities of the land, which he described as the "most beautiful region in the world." Louis Hennepin disputed wild claims that the land would yield a double crop of corn each season—he believed that three or even four crops a year were possible from "vast meadows, which need not to be grubb'd up, but are ready for the Plow and Seed." He also remarked that "certainly the soil must be very fruitful, since Beans grow naturally without any Culture." Louis Armand de Lahontan described the soil as "so fertile that it produces (in a manner without Agriculture) our European Corn, Pease, Beans, and several other Fruits that are not known in France." Bacqueville de La Potherie admired the Native Americans, "who are in possession of the most beautiful region that can be seen [anywhere]." When Jean Cavelier attempted to persuade a group of them to accompany him north, they refused. He recorded that the local inhabitants, "being in the most fertile, healthy and peaceful country in the world, they would be devoid of sense to leave it and expose themselves to be tomahawked . . . or burnt [by enemies] on their way to another, where the winter was insufferably cold, the summer without game, and ever in war." Even those few early white explorers who were less effusive about the Illinois country admitted that while some Illinois lands "require Toil and Trouble . . . they can sufficiently recompense in a little time, those who will be at the pains to cultivate them." Such praise from the earliest observers of the new land appeared so regularly in official reports to secular or religious superiors and in correspondence to intimate friends that it must be considered an accurate reflection of their emotions upon first embracing the Illinois country.[8]

The French wasted little time in claiming the territory. By the late 1600s, they had established posts along the major rivers and lakes in the Illinois region to trade for furs with the Native Americans. During the eighteenth century, however, the land changed hands as a consequence of international conflicts. The French claimed the area in the first half of the eighteenth century and built a series of forts on their Mississippi frontier to defend their title. The Treaty of Paris, in 1763, awarded the Illinois territory to Britain. In 1778, George Rogers Clark defeated the British garrisons at Kaskaskia and Vincennes and delivered the Illinois region into the hands of the new American government. For the next five years, Clark and his men fought against the British and their Native allies to defend the Illinois and Kentucky territory. Illinois was organized as a territory of the United States in 1809. During the War of 1812, many tribes who believed they had been betrayed by the federal government or suffered at the hands of American settlers allied with the British and engaged in scattered fighting across northern Illinois. At one point they even attacked Fort Dearborn, the present-day Chicago, and killed or captured the 100 soldiers and settlers who had taken refuge there. During the conflict, according to local DeKalb County lore, Captain Webb rode from Fort Dearborn, through St. Charles, to a post on the Mississippi River, becoming the first white to see the land that would become DeKalb County.[9]

The British withdrawal from the United States resulted in a tentative, uneasy truce between the Native Americans and white settlers on the frontier of northern Illinois. The newcomers feared raids, and the Native Americans were suspicious of whites who encroached on their traditional lands. Moreover, despite the fact that the prairie soil was rich beyond belief to farmers accustomed to farming the thin, stony soils in New England or parts of the trans-Appalachia west, it was not easily farmed with conventional tools and techniques. For pioneering farmers who were accustomed to hacking their way through unbroken forests and carving out woodland farmsteads, the relentless expanse of open prairie was a foreign and foreboding place to begin a new life. Even when newcomers succeeded in raising crops, the isolation of the region, lack of transportation, and sparse population made it difficult to market their agricultural products. Consequently, white movement to the DeKalb area was slow. A sufficient number of people settled in the Illinois territory, however, so that Illinois was admitted as the twenty-first state on December 3, 1818. By the late 1820s, increased white settlement on Native American land in the northern parts of the new state further heightened tensions between the local Sauk and Fox peoples and the settlers.

To relieve this dangerous tension and open more land for white settlement, in 1829 the federal government ordered the Sauk and Fox to leave their homes in northern Illinois. Most bands moved north or west, across the Mississippi River.[10] Whites occupied vacated villages, planted abandoned fields, and desecrated Native American sacred sites. In 1832, the Sauks' chief spokesman and war leader, Makataimeshekiakiak, or Black Hawk, a veteran of the War of 1812, returned with some of his tribe across the Mississippi River into Illinois to establish a new village and begin farming. Howls of protest went up from area settlers and politicians, who complained that the Sauk "threatened our lives if we attempt to plant corn, and say that we have stolen their land from them, and they are determined to exterminate us." Such pleas compelled President Andrew Jackson, already hostile to Native American interests, to send federal troops into action to force the Sauk back across the river. By late 1832, United States regulars and Illinois militia led by General Winfield Scott pursued Black Hawk and his followers north, up the Rock River valley into Wisconsin, and then west across the Mississippi. At the mouth of the Bad Axe River, in Prairie du Chien, the soldiers attacked them, killing many of them and ending their attempt to regain their traditional homelands in Illinois. During the conflict, a small force of soldiers left Fort Dearborn in pursuit of Sauk and Fox raiders. They crossed the Fox River near present-day St. Charles and camped along the banks of the Sycamore River, later renamed the Kishwaukee.

A year later, in 1833, after Black Hawk's defeat, white prospectors and hunters briefly explored what was to become the southern part of

DeKalb County. When they encountered small Native American hunting parties enraged by their defeat and the government's policies, the whites quickly retreated back to their homes in Ottawa, Illinois. The tide of settlement, however, could not be turned back by angry but isolated bands of hunters. Farmers established permanent settlements in counties surrounding DeKalb, including neighboring Lee, Ogle, Kendall, and DuPage counties. During the spring and summer of 1834, more whites came to the area of DeKalb County with an eye toward taking land for farming and settlement. In general, the earliest settlers came from western New York, New England, northern Ohio, and a handful from the South.[11] Prices for farmland in these areas were increasing during the 1830s, forcing many to look to the relatively low-priced, newly opened midwestern lands as a more economical and profitable alternative. The first permanent white settler in DeKalb County was John Sebree, who settled in the area in 1834. Sebree, a Virginian who had already farmed in Indiana before moving to Illinois, spent his first winter in an abandoned wigwam and subsequently erected one of the first permanent buildings in the county, near the town of Squaw Grove (currently known as Hinkley). Other farmers, such as Marshall Stark, Ruben Root, and Frederick Decatur Love, returned to the area within a year or two of their initial reconnaissance and became some of the county's leading citizens.[12]

Early explorers, at a loss to describe objectively the wild prairie lands beyond the fringe of white settlements, could not help injecting emotions into their letters and dispatches. Later chroniclers turned increasingly to literary metaphors to convey the vastness of the land. They relied upon themes of romance, destiny, the promise of fortune, an ennobled spirit, unsurpassed natural beauty, and personal liberty to portray the new land. During the late 1790s, a French traveler, Victor Collot, wrote poetically, "The Province of Illinois is perhaps the only spot respecting which travelers have given no exaggerated accounts; it is superior to any description which has been made, for local beauty, fertility, climate, and the means of every kind which nature has lavished upon it for the facility of commerce."[13] Gilbert Imlay described "the wild regions [of Illinois], where the sweetened breezes attune the soul to love." For another writer, the prairie was "a sort of flat paradise." Illinois, according to H. Cogswell Knight, was "a Good poor man's country; for, in truth, the clusters are clusters of Eshcol, the land foams with creamy milk, and the hollow trees trickle with wild honey." "With a soil more fertile than human agriculture has yet tilled;" wrote C. W. Dana, "with a climate balmy and healthful, such as no other land in other zones can claim; with facilities for internal communication which outrival the world in extent and grandeur—it does indeed present to the nations a land where the wildest dreamer on the future of our race may one day see actualized a destiny far outreaching in splendor his most gorgeous visions." The American poet William Cullen Bryant wrote of his visit to Princeton, Illinois, in 1833:

> These are the gardens of the Desert, these
> The unshorne fields, boundless and beautiful
> For which the speech of England has no name—
> The Prairies, I behold them the first,
> And my heart swells while the dilated sight
> Takes in the encircling vastness.[14]

Few early white settlers arriving in northern Illinois during the 1830s, however, shared such sentimental visions of the new land. Instead, families who moved to DeKalb County entered an alien landscape that was far removed from experiences of most contemporary Americans. The prairie was awe inspiring and intimidating in its sheer emptiness. The softly waving grass and gently rolling landscape offered early travelers the illusion of a boundless, green ocean. Indeed, contemporary authors often turned to nautical metaphors when describing the prairie, for nothing else seemed adequate to portray the magnitude of the empty land to those who had not experienced it firsthand, and nothing else was as well suited to capture its foreboding mystery. "Looking toward the setting sun, there lay, stretched before my view, a vast expanse of level ground. . . . There it lay, a tranquil sea or lake without water . . . with the day going down upon it . . . and solitude and silence reigning paramount around," wrote Charles Dickens about his visit to the Illinois prairie in 1842. "For miles we saw nothing but a vast prairie of what can compare to nothing else but the ocean itself," wrote *Boston Courier* journalist J. H. Buckingham about his trip with Abraham Lincoln across Illinois in 1847. "The tall grass, interspersed occasionally with fields of corn, looked like the deep sea; it seemed as if we were out of sight of land, for no house, no barn, no tree was visible, and the horizon presented the rolling of waves in the afar-off distance." Bryant noted simply that "the green heights and hollows and plains blend so softly and gently with one another." Illinois booster Frederic Gerhard warned his readers, as late as 1857, that "[t]hose who have not seen the prairie should not imagine it like a cultivated meadow, but rather as a heaving sea of tall herbs and plants, decking it with every variety of color." Morris Birkbeck, in his *Letters from Illinois,* warned that "[e]migration to the extreme limits of this western America will not repair a bad character . . . the wild lands of Illinois will yield no repose to perturbed spirits." Some who saw the DeKalb prairie must have believed that during their lifetimes it could never be successfully settled. Most who arrived, however, overcame their fears, worked hard to master the prairie, and hoped that the land would yield its agricultural treasure.[15]

DeKalb County was created in March 1837, and in the following decade, its population expanded dramatically despite both spiritual and practical farming challenges.[16] During the 1840s, DeKalb County's population continued to increase. Immigrants and the native born moved into the area, overland from frontier towns like Chicago and St. Charles or up the Illinois, Rock, and Fox rivers from population centers further south, to take up new agricultural land. When the county was first organized, there were only a few hundred people living in the region. In 1840, the county population had already grown to 1,697.

One of the greatest limiting factors in the settlement of DeKalb County was transportation. The Kishwaukee River was far too small for barge transportation of heavy agricultural cargos. Consequently, settlers looked to roads as the most direct links to markets. Regular stage service between Chicago and Ottawa and between Chicago and Galena began in 1834. In 1840 roads opened between St. Charles and Dixon and between St. Charles and Sycamore. A year later, a north-south road opened between Ottawa, Illinois, and Beloit, Wisconsin, that ran through the heart of the county.[17] Even by the primitive standards of the day, these roads were terrible. They were barely passable under the best circumstances. Wagons sank up to their boxes on the unimproved roads built through the heavy loam soil. Animals struggled to pull heavy loads through the sticky mud or across streams. Wagons and draft animals broke down. Entrepreneurs built a few wooden plank roads, a primitive form of paving, but they were expensive to construct, deteriorated quickly, and were difficult to drive upon. They warped, rotted, and disintegrated after only a few seasons of traffic. Whatever their shortcomings, these roads offered newcomers some access to the interior of northern Illinois, and, perhaps as important, the roads connected the settlers with the outside world.

DeKalb County Townships

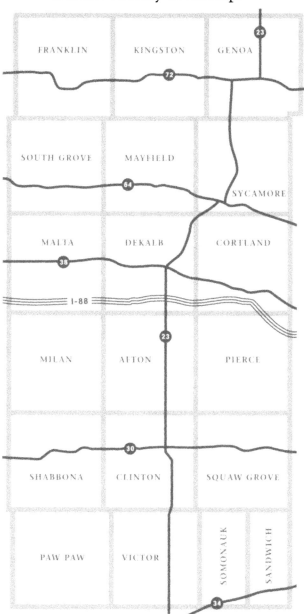

The population also increased when the federal government offered much of the land in the central part of the county for sale. The sale was typical of many chaotic government land offerings, and, because much of the land was already occupied, special care had to be taken during the auction to ensure that the land went to those who had the greatest stake in the outcome. "The land was sold at auction," wrote nineteenth-century historians Jesse C. Kellogg and Henry Lansom Boies, "and from each neighborhood one trusty man was selected to bid off the property, as it was offered, while the remainder of the crowd stood around, armed with clubs, ready to knock down any spectator who should dare to bid for lands that had been claimed and occupied by any of their party."[18]

For early settlers accustomed to forested farmlands farther east, the most attractive places in the county for them to locate their new homes were the timber groves that snaked along the county's few streams. Consequently, the areas that appealed to the earliest settlers were in what is now Shabbona and Paw Paw townships, in the southwestern portion of the county. These areas were originally wooded, providing timber for buildings and fences—at one time, there were eight sawmills along the banks of the Kishwaukee River, though they quickly depleted the timber resources and went out of business. The open prairie townships north of Shabbona and Paw Paw remained thinly settled, but, as the county population increased, more settlers turned to those areas for land. While the lush prairie soil was ideal for growing crops, the groves provided settlers with timber for homes, barns, fences, and firewood. The trees also gave shade in the warm summers and some protection from the harsh winter winds. The streams provided water for families and livestock. Sebree, for example, chose land that included forty acres, with plenty of timber, along a stream. More important, however, the soils along the streams were generally lighter timber soils that were easier to plow with the cast iron and iron-shod wooden plows that were common at the time—to this day, the lighter timber soils of northern DeKalb County support different crops than the heavier, denser soils to the south. These plows, perfected in the looser and thinner soils of the eastern states, performed poorly in the deep soils of northern Illinois. Plowmen who attempted to "break" the virgin prairies were forced constantly to sharpen and clean their plows, sometimes every few steps. In 1837, however, at his blacksmith shop in the town of Grand Detour in neighboring Ogle County, John Deere perfected a lightweight plow with a highly polished steel moldboard and share that sliced through the prairie soil. Ten years later, Deere was manufacturing a thousand "self-polishing" plows a year, making it possible for farmers to settle successfully in what was then the "West."[19]

During the first half of the nineteenth century, such new technologies, combined with the promise of the fertile soil, attracted people to DeKalb and the county population grew. In 1850, the county population was 7,540, most of whom lived on farmsteads. That same year, the city of DeKalb, today the county's largest city, had a population of 486. The population of other communities was even less. Most town residents worked the land, either as landowners or laborers, at least some time during the year. The local business sector was small, but the town supported a blacksmith, two general stores, and a tavern. Rural folk also found they could make a modest living. There were, however, structural impediments to economic success for area residents. DeKalb was located in the interior of the state and had little access to ground or water transportation. Consequently, there was little trade between neighboring communities. More important, however, DeKalb farmers found that the high cost of moving heavy grain surpluses by wagon over poor roads, or driving livestock, to mills or docks on the Rock or Fox rivers or the Great Lakes made it nearly impossible to expand their output beyond what could be locally consumed. The arrival of the railroad to St. Charles in the late 1840s and the construction of a plank road between Sycamore and St. Charles eased the transportation problems somewhat, but the county remained largely isolated. Periodic droughts during the 1840s ruined wheat crops, the predominant agricultural commodity of the region at the time. In 1842, the county's State Bank failed, undercutting the region's economy further. There were few buyers for land, and many farmers found themselves land-rich but poor in every other way. Costs for goods remained high. Taxes, land purchases, mortgages, and store purchases demanded cash, but the scarcity of capital on the margins of white settlement meant that agriculture yielded the farmers of the era only modest returns and kept them largely at a subsistence level of farming.[20]

In 1853, however, after careful consideration of several factors, including ease of construction, maintenance costs, and profitability, the Illinois Central Railroad laid tracks through DeKalb County, linking the county to the rest of the nation and bringing new economic opportunities to the area. The new rail connection made the old plank roads irrelevant for anything but local traffic. The influx of population and money into the region as a consequence of the railroad permanently changed the community. As often happened when a new railroad arrived in an underdeveloped region, DeKalb experienced a boom in population, jobs, wages, and general prosperity. The railroad employed around 10,000 workers statewide in various jobs, including construction, and perhaps as many as 100,000 men during one five-year period during the 1850s.

Many of these laborers were temporary workers who boarded in the local communities during construction, bought food and other supplies, and then moved on to new construction sites or jobs. The railroad also fueled more long-term changes. It hired men to remain in the area permanently to operate and maintain the new track and run the stations and other railroad facilities in the communities it served. Some construction laborers were farmers working to raise a financial stake to begin, or renew, farming and were actively looking for places to purchase land. A few of them stayed in DeKalb County after the railroad construction gangs moved on. Other laborers, attracted by the region's financial opportunities, also settled in DeKalb County. These newcomers and their families dramatically increased the county's population, which grew to over 19,000, a 70 percent increase between 1855 and 1860. The city of Sycamore, the county seat, grew 72 percent in those five years, and DeKalb had 1,900 people in 1860, more than triple its size from the 1850 census.

The newcomers had good reason to remain behind. Area farmers, eager to expand production and profits promised by regular rail service, hired more farmhands to work greater amounts of land. Demand for labor kept wages up in the seasons before the advent of large-scale industrialized farming after the Civil War. Living conditions in the area were often better than those laborers had endured on marginal eastern farms or in the teeming eastern or European cities, and their wages were usually much higher. New service providers, including skilled laborers, businessmen, and even a lawyer and doctors, catered to the growing rural and urban population. Merchants and manufacturers relocated in DeKalb County's towns and villages. In 1854, H. A. Hough published the *Republican Sentinel,* DeKalb County's first newspaper. Higher prosperity generated greater tax revenues for local improvements, including an elegant new brick elementary school completed in 1860 that was among the finest in the state. Shopkeepers relied on the railroad to deliver their goods cheaply and efficiently. In addition to basic supplies, such as fabric, manufactured farming equipment, and household goods, luxury goods began to appear more regularly on store shelves. Railroads delivered heavy raw materials, such as iron and steel for horseshoes and wheels, to local craftsmen. The agricultural sector especially benefited from the transportation services provided by the Illinois Central. Area farmers now had the means to market their harvests successfully beyond the local area. Production and profits soared. In 1850, DeKalb County farmers produced about 216,000 bushels of "Indian corn"; ten years later they produced 496,000 bushels. Farmers sold much of the increase to markets in Chicago or St. Louis, or fed the grain to livestock, which, in turn, they sold to butchers and early packinghouses in the region. The railroad also brought new goods and agricultural innovations to the farmers that further boosted production and improved living standards. Fast and reliable freight services standardized and regularized delivery and transportation times and costs, making it easier for farmers to predict their costs and maximize their profits. The railroad was probably the most important technological innovation and engine of prosperity in DeKalb County during the nineteenth century.[21]

The completion of the railroad into DeKalb County dramatically increased the prosperity of the region and offered greater financial security to area farmers. Corn, which had been grown in the area by Native Americans and later by white settlers, was more dependable and profitable in most of the county and supplanted wheat and oats as the principal crop throughout the county except in the northern part, where the soil was thinner. Property values rose. The new affluence also led county residents, most of whom were Methodist and Lutheran Protestants, to be more conservative. Political records from the early 1850s suggest that a majority of residents were Democratic, but there were strong minorities of Free-Soilers and Whigs. When the Republican Party emerged in 1854, the Free-Soilers and Whigs eagerly joined it, as did increasing numbers of Democrats, particularly those who were uncomfortable with the Democratic Party's Southern ties. Many in the county were especially attracted by the Republican Party's antislavery stance, and several prominent county residents were vocal abolitionist sympathizers. A few of them, like David West, Ira Nichols, and Joshua Townsend, were actively engaged in the Underground Railroad. As national political tensions escalated, DeKalb men increasingly stood by the Republican Party until, by the end of the decade, DeKalb County was firmly entrenched in the Lincolnesque mode of Republican politics—during the 1860 election, Lincoln received 3,049 votes to Douglas's 950, and four years later Lincoln won 2,985 votes to McClellan's 741.

After the fall of Fort Sumter in 1861, DeKalb County energetically supported the Union cause. By October, there were already nine companies of DeKalb men in Union service. Ultimately, about 3,000 DeKalb men fought in the war, and until 1864, no draft was necessary to fill the DeKalb County recruitment quotas. They fought in many campaigns across the nation, including the Virginia Peninsula, Gettysburg, Fort Donelson, Vicksburg, Mobile, and Sherman's Georgia campaign. Horace W. Fay, a longtime resident of Squaw Grove and the county surveyor, became a captain and chaplin in one of the Union's Black regiments before he died at Vicksburg. Despite the initial wartime economic panic, patriotic citizens raised $30,000 to support the families of the volunteers. The relentless call for troops and the bounties paid for enlistments drained county coffers and increased tax assessments, but, by 1863, the worst of the economic crisis had passed. Wartime demand for agricultural products soared, and for the remainder of the war the county prospered.

During the postwar decades, DeKalb County continued to flourish, and agriculture remained the most important economic activity in the county. Although farmers in DeKalb endured periodic agricultural boom and bust cycles, the fertility of the county land and its proximity to major midwestern markets meant that DeKalb farmers avoided the worst consequences of the periodic economic downturns. As the century wore on, business and industry increasingly emerged as an important component of the county's prosperity. Many of the small businesses, such as the retail shops and service companies, served their local communities. The county had six major banks by 1900. Other businesses produced goods for a regional and even national market. Just before the Civil War, for example, the Marsh brothers began building their Marsh harvester in Plano, Illinois. After the war, the Marshes, and other implement manufacturers, built plants in Sycamore and DeKalb, and soon the county became an important source of a broad range of farm implements. Other local industries also contributed to the county economy, including a brass foundry, an electrical wire and cable company, a dairy package manufacturer, a cannery, and even a cigar manufacturer.

The most important industry during the late 1800s, however, was barbed wire. As farmers encroached on the cattle ranges, roving cattle herds frequently destroyed crops. Cattle ranchers also found it necessary to fence out competitors. In much of the West, there were few trees for settlers to build wood fences, board and rail fences were prohibitively expensive, and cattle ignored standard wire fences. During 1873, Joseph Glidden perfected a form of barbed wire that discouraged cattle from breaking the wire fence, and in 1874 the United States Patent Office issued Glidden patent number 157,142 for his invention. A year later, Glidden and his partner Isaac Ellwood opened a factory in DeKalb to manufacture their new invention. Jacob Haish also opened a factory to produce his own design of barbed wire. The expanding American network of railroads carried spools of plain wire from wire mills, such as the Washborn and Moen mill in Worcester, Massachusetts, to the barbed wire factories in DeKalb. The railroads then carried the barbed wire, and the DeKalb wire kings' advertising, to buyers throughout the American West—DeKalb became internationally famous as the home of barbed wire. Even today, historians agree that barbed wire was one of the most important inventions in the development of American agriculture during the nineteenth century. By the turn of the century, the "Big Three" of Glidden, Ellwood, and Haish had diversified commercial interests throughout the region, were important philanthropists, and were the leading citizens of the county.

The combination of increasingly mechanized agriculture and a strong industrial base transformed DeKalb during the late 1800s. By the turn of the century, only a handful of the oldest farmers could remember the agriculture ordeals of the 1830s and 1840s. The Populist revolution that swept the American heartland during the 1890s attracted few adherents among DeKalb's prosperous, conservative, moralistic, and staunchly Republican farmers. Economic success of the county was reflected, instead, in people's commitment to local institutions. By 1900, the county boasted more than sixty churches, most of them Protestant. Methodist and Lutheran congregations were the most common. Nearly every township and village had its own newspaper. Some, such as the Somonauk *Free Press*, the Sandwich *Prairie Home and Advertiser*, and the *Genoa Siftings*, lasted only a few years. The *DeKalb Daily Chronicle*, on the other hand, first appeared in 1879 and remains the most important journalistic voice in the county. County newspaper publishers were usually civic-minded, and their papers—more than thirty journals were established in the county during the second half of the nineteenth century—played an important role in delivering information and entertainment, influencing public tastes, supporting local projects, policing morals, and building a sense of collective identity that helped bind readers together as a community. The first DeKalb County schools began in 1837, and by the end of the nineteenth century, small elementary schools dotted the county, while the larger communities, Sycamore and DeKalb, also had high schools. The city of Sycamore began its quest for a public library in 1875, and the city council began supporting it officially in 1891. DeKalb created its public library in 1893, and Sandwich did so in 1898. In 1892, Clinton Rosette, editor of the *DeKalb Daily Chronicle* and member of the board of trustees for the normal schools in

Illinois, began his crusade to locate a normal school in northern Illinois. Three years later, the Illinois legislature agreed to build a new "teachers' college" to serve the northern Illinois region. Competition for the school was intense between city boosters in Dixon, Freeport, Polo, Rockford, and DeKalb. With the help of the city and the "Big Three," who offered to donate land and money for the new institution, and local residents, who convinced the visiting board that their town was the best location, however, the new Northern Illinois State Normal School found a home in DeKalb. The first session for the NISNS began in September 1898 in the nearly-complete college "castle" building, with 16 faculty and 139 students.

By the opening years of the twentieth century, DeKalb County was typical of hundreds of successful agricultural counties along a broad sweep of America's midwestern corn belt. The county only vaguely resembled the prairie wilderness that greeted, and challenged, the earliest white settlers. The sea of grass had long since been turned into farms producing crops and large livestock herds. European immigration into the county during the nineteenth century resulted in a rich cultural diversity. A network of improved roads crossed the county, and, though they were not always easy to navigate in foul weather, they bound even the most remote county farms to the outside world. The railroad linked DeKalb County to the rest of the nation. The few rough pioneer cabins and barns that remained were crumbling, picturesque reminders of a pioneer past that was remembered only by the old-timers who now lived on comfortable and prosperous farmsteads or in neat town homes. Large barns, grain cribs, livestock pens, and maturing windbreaks reflected the agricultural affluence of the area. The wealthy county elites built mansions along tree-lined streets in the county's towns and villages. Most communities had thriving commercial blocks and industrial sections, and even the smallest hamlets had a couple of two-story, false-front buildings housing businesses that met the everyday needs of local customers. In 1905, the county completed its new courthouse in Sycamore, a three-story, neoclassical Victorian structure, hailed at its opening as a "temple of American architecture." "There is not a building in the state," the *Sycamore Prospectus* noted with obvious civic pride, "that is better adapted, more substantially built, or furnished in better taste."[22] Rural and small-town folk never considered themselves living on the margin of American civilization. Rather, they believed that they were modern Americans and ideal citizens of the nation who took pride in their schools, churches, clubs, and other social organizations that bound them together into close-knit communities.

In other ways, however, DeKalb County was unique among corn belt counties during the opening years of the twentieth century. Its soil was some of the most productive in the world, and its farmers enjoyed a standard of living that was the envy of the entire American agriculture sector. It was the home of barbed wire and other important industrial and commercial enterprises that enriched the county. Its proximity to Chicago and Rockford and its extensive rail connections meant that the county was bound to regional, national, and international markets. The new Northern Illinois State Normal School ensured that it would be a center for learning in the region. Perhaps most impressive of all, however, was that the county was home to a handful of progressive DeKalb farmers, bankers, newspapermen, and civic-minded community leaders who were farsighted enough to consider the long-term future of their community. Together, they created an organization dedicated to preserving the vitality of the county in which they made their homes.[23]

1 A New Era

THE ROOTS OF THE
FARM BUREAU MOVEMENT

• On Saturday, July 8, 1911, more than a thousand local farmers and townsfolk gathered at the Whitmore-Oakland farm, between Sycamore and DeKalb, Illinois, to celebrate the summer season. It was a festival not much different from agricultural holidays people have enjoyed throughout history, and in many respects predominantly rural DeKalb County appeared to be a place more representative of the nineteenth century than the twentieth. Like most American farmers of the time, DeKalb farmers embodied Calvinist ideals that stressed hard work, self-reliance, thrift, and sobriety. The 1910 U.S. census recorded that DeKalb County had 2,481 farms, divided about equally between owners and tenant farmers, that accounted for more than 95 percent of the county's land. Farm prices averaged between $100 and $125 per acre and were among the highest in Illinois. The census also showed 33,457 residents living in DeKalb County, of whom about 60 percent lived on farms or in communities with fewer than 2,500 residents and were therefore classified in the census as "rural." Although there were several important industries and manufacturing companies in the county, these 1910 demographics are more reflective of an earlier period in American history, when farming was the dominant economic enterprise in the nation. Even the county's urbanites were not far removed from the farm and lived in communities that depended directly or indirectly on agriculture. At the fair that afternoon, the men spoke together about families, animals, business, machines, markets, the weather, and of course, corn— corn, after all, was emerging as northern Illinois's most dependable and profitable commodity. The women, too, shared news with friends and family while they managed the day's picnic food and watched children.

Most dreamed as much as their husbands about the future, and they worked as hard as their men to make their farms successful.[1]

Yet, despite the traditional atmosphere, the gathering was very modern. In an impromptu, and dangerous, demonstration of old and new technology, Charlie Broughton hitched his Buick automobile to Mr. Hammond's late-model Marsh harvester, a machine built to work at the pace of draft horses. The car "walked away with the harvester as if it were nothing," wrote one reporter, swamped the men who were filling and binding the grain sacks, and threw the men off the machine's work platforms when Charlie rounded the first turn at high speed. It was the comical highlight of the afternoon. Few spectators, however, were really surprised by the outcome. Evidence that the modern world was succeeding the traditional one was obvious throughout DeKalb County. The fertile prairie ecosystem that had attracted the earliest white settlers to the region was already converted to farmland and livestock pens by the early 1860s. Many county locations and geological features retained their Native American names, but the conflict and the forced removal of the tribes decades earlier was a vague and distant memory. The pain of the Civil War was receding into the past to such a degree that area residents erected a huge statue on the courthouse square in 1896 to honor the veterans and remind future generations about the sacrifices made by their fathers, uncles, and grandfathers on America's killing fields. Only a handful of county residents still spoke their native languages more comfortably than English.[2]

Old-timers, such as County Treasurer Ed Johnson and Sycamore attorney William Allen, entertained the curious crowd at the fair with demonstrations of antiquated scythes, sickles, and cradles. Younger farmers at the celebration, however, had never experienced firsthand the backbreaking labor of manual harvest. Instead, they depended upon harvesting machines that boosted farm yields beyond their grandparents' imaginations. A self-binding harvester at the picnic, for example, cleared a path through the grain in a fraction of the time that it took the experienced human cradlers. Although there were still plenty of horses and carts at the festival, machines were rapidly displacing animals as the chief source of power on farms. The huge, and powerful, but slow-moving, cumbersome, and smoking steam engines were already in operation on several DeKalb County farms as early as 1905, but lightweight, fast, and efficient gasoline engines were the wave of the future. "The faithful horse is being still further supplanted," reported the *Sycamore True Republican* about the new tractors in 1909. "Now it is his old job of plowing that is being taken away from him—and if he understood it, he would feel grateful that the ingenuity of man is depriving him of that hard work. The time is passing away when the faithful horse hitched two, three, or four abreast, will be obliged day after day to strain himself in turning over the heavy soil." The pioneer in gasoline tractors in DeKalb County was Frederick B. Townsend of Sycamore, who had a twenty-two-horsepower gasoline tractor working on his farm south of Sycamore in 1908. M. H. Coleman and George Dick, from the Mayfield area, each bought twenty-horsepower International Harvester gasoline tractors the next year for their John Deere gang plows. Gasoline tractor demonstrations, in towns like Wheatland and Big Rock, attracted interested farmers, many of whom placed orders for the new rigs. A gasoline tractor was not inexpensive—an outfit including plows cost as much as $2,000—but it had important advantages over steam.[3] It carried its own fuel, had no need for water, had fewer gauges to watch, required only one operator, and was simpler and more reliable. At fifteen cents per gallon, gasoline for a day's work cost about $2.50, less than the cost of feeding enough horses to accomplish the same work. Moreover, unlike horse teams, tractors did not require attention when the workday was finished.[4]

Area residents also relied on the railroad and a network of gravel and hard-surfaced roads, including the newly completed Lincoln Highway that ran through DeKalb and the northern part of DeKalb County, to carry their produce to market and deliver the necessities of modern life into north-central Illinois. Towns like Sycamore and DeKalb were also manufacturing centers that had produced such goods as barbed wire and farm implements for decades. The Clover Dairy Farm, owned by the Gurler brothers, produced sanitary dairy products for northern Illinois and Iowa in their state-of-the-art facility. So pure was their output that in 1900 the brothers shipped fresh milk and cream by railroad and steamship from DeKalb to the Paris Exposition, yet, after seventeen days in transit, the milk was still fresh upon arrival. Several large and small newspapers in the county reported on local affairs and published news for farmers who were aware that national and international events might influence their financial success. The prosperous DeKalb farmers lived in a modern agricultural sphere dominated by carefully selected seed, purebred livestock, laborsaving technology, sophisticated communications, industry, rapid transportation, regional credit and banking relationships, and national and international markets. In the summer of 1911, DeKalb County was thoroughly up-to-date, and its residents intended it to remain that way.[5]

They had every reason to be confident about their futures. The rough pioneer days were long past. Nationally, farm incomes rose steadily, though not dramatically, every year between 1897 and 1910. During the "Golden Age" of American agriculture, the years between 1909 and 1914, economic conditions were so favorable for farmers that the period is still used as the basis for farm parity price calculations.[6] Technological developments, especially the gasoline tractor, led to greater mechanization of farming, a process that reduced the physical hardships of the pioneering days while dramatically improving farm efficiency. Millions of acres previously used for the production of grain and forage for animals were freed for production for human consumption. The world was at peace, and industrialized and urban populations at home and abroad demanded farm products. Farm output increased at a rate slightly slower than that of consumer demand, and production costs remained generally stable, resulting in relatively high produce prices and profits. The total value of agricultural exports continued to increase during the period, and tariffs and international trade terms usually favored American farmers. Modern farming required access to capital to finance new machinery, farm improvements, and the consumer products that were fast becoming mainstays of the American standard of living. Under such favorable economic conditions, credit was easy to find. Property values were rising, and consequently farmers' net worth, in large measure because the supply of good-quality farmland was decreasing nationally. Substantial long-term improvements made to existing farms increased property values further. The overall prosperity of the nation and better organization of financial markets made borrowing easier, while stability of farm values made farm loans more attractive for lenders in prosperous areas such as DeKalb County. Throughout the period, farmer equity generally outpaced mortgage debt. Certainly, the financial risks were great and farmers remained at the mercy of natural and market conditions beyond their control, but the financial rewards for farmers who mastered new agricultural business skills could be spectacular. "There has never been a time," noted the Country Life Commission's Report to the Senate in 1909, "when the American farmer was as well-off as he is today, when we consider not only his earning power, but the comforts and advantages he may secure."[7] Midwestern farmers, especially, were making more money and living better than ever before.[8]

Nevertheless, during this period of unparalleled agricultural prosperity, American farmers also experienced uncomfortable changes. Farm incomes were on the rise, but not as quickly as those in other sectors of the economy. Although DeKalb County farms prospered, agricultural wealth was not shared equally among the nation's farmers. In several regions, especially the South, rural poverty was a widespread problem. Easily cultivated new agricultural lands were largely gone by the turn of the twentieth century. As inexpensive land disappeared and farm values appreciated, costs and opportunities for new farmers, agricultural laborers, and tenants to acquire their own farms dwindled. Farm tenancy was rising; in 1910, 37 percent of all farm operators were tenants. Although real property equity and prices remained high and credit terms were favorable, farmers nevertheless carried considerable amounts of mortgage debt. Memories of periodic agricultural depressions during the nineteenth century, the most devastating occurring during the mid-1890s, made farmers uneasy about their futures. Science and technology made farming more predictable and potentially profitable, but marketing strategies and farming practices rooted in the nineteenth century too often remained inefficient and uncertain. There was also some truth to farmers' claims that monopolistic business practices, discriminatory transportation rates, and exploitation by processors and middlemen eroded their profits.

Socially, cosmopolitan values and conventions challenged rural ones. Country folk gradually abandoned older ways of living and adopted urban expectations. Mechanization of agriculture forced rural laborers into industrial and service jobs. Cities promised, and often delivered, greater social and intellectual opportunities than did heartland villages and towns. In the country, the lack of adequate social services, including schools, as well as a perception among many rural folk that farm life was sterile and isolated, also made city life seem appealing. In rural areas roads were often poor, local medical services were meager, and sanitation was primitive. The best-educated rural professionals often turned to the cities as more comfortable places to live, thereby jeopardizing the quality of all kinds of rural services. Nationally, agriculture was more profitable than it had been in earlier decades, and farmers' long-term land equity was generally increasing, but many farmers were simply not making as much money as quickly as people in urban occupations. These difficult conditions did not appear suddenly and were not universal, but all were features of modern agriculture life during the early twentieth century.[9]

Of all the difficulties confronting agriculture, however, the exodus of rural Americans to towns and cities and the resulting erosion of rural society were especially alarming for Progressive-era reformers during the opening years of the twentieth century. Since the eighteenth century, Americans believed in a Jeffersonian ideal that viewed farmers as the foundation of American democracy. Noble and mythical notions about the inherent importance of agriculture remained strongly rooted in the national consciousness. Industrialization during the nineteenth century, however, gradually shifted the real economic, political, and social power of the nation away from the country and to the cities. It successfully challenged the ideas that agriculture was the most important basic economic activity of the nation and all other enterprises were somehow less important. Yet, rather than erase romantic notions about agriculture, the rapid modernization of America at the turn of the twentieth century reinforced them. Ironically, some of the most vocal support for the agriculture ideal came from urban reformers. For many, utopian agrarian life was a standard by which to criticize the social and economic problems of the cities caused by industrialization. Many urban Americans came to believe that the movement of farmers into the cities threatened to destroy the pastoral underpinning of the nation. They feared losing their moral, agrarian anchors to a future dominated by what they saw as ambiguous, shifting, relative, and alien values of the cities. Agrarian life and values were the cure for the social and political infirmities that seemed to have infected the nation. Farmers were archetypal Americans, and American civilization itself was at stake if the agricultural population continued to decline.[10]

Some Progressives believed, often accurately, that the root of this problem was the inadequacy and backwardness of country living and proposed sweeping changes to improve rural life and reverse the rural population decline. This new reform outlook was different from earlier agrarian crusades. During the nineteenth century, backward-looking rural reformers assumed that farmers were "the people" of the nation. They tended to see the national issues in absolute terms—Americans either shared the interests and values of farmers because they were dependent upon agriculture directly or indirectly, or they were not true Americans. Consequently, the focus of earlier agricultural reform movements, such as Populism, was to recover what farmers saw as their traditional importance in American society that had been usurped by antidemocratic, antiagrarian elements that were corroding the nation's soul. By rejecting modernity, farmers could retain their historical status and realign the national conscience.

Unlike these nineteenth-century agricultural reform movements, which were largely futile protests against the declining political and economic status of farmers, the new century's Progressive reformers sought to ennoble farmers and encouraged them to assert their rightful place in society within the context of modern America. Farmers could no longer simply try to rewind the clock and turn their backs on the changes that were taking place around them. They needed to embrace, and master, the techniques and values of the modern world if they wished to regain their historical importance in society. It was a Progressive reform impulse—deeply rooted in rural nostalgia, humanitarianism, and economic self-interest utilizing modern scientific knowledge, technological expertise, and bureaucratic organization—to encourage farmers to pursue power and thereby reestablish themselves as vital participants in the economic and social fabric of the nation.[11]

Ironically, the first to warn of the problems in the country and suggest a fundamental reassessment of agricultural life were hardly model Jeffersonian yeomen. Leadership came instead from a wealthy class of farmers who embraced contemporary commercial agriculture and business values. Support also came from publishers and editors of local and regional farm newspapers and journals who zealously endorsed improved farming and management techniques. A new cadre of academics at land-grant colleges and bureaucratic professionals in federal and state departments of agriculture advocated scientific and technological improvements. Commercial groups whose financial fortunes were closely allied with agriculture, such as banks, mail-order houses, railroads, merchants and chambers of commerce, and implement dealers also supported broad reform of rural life. Rural clergy supported reform for spiritual and moral reasons. Still others, including teachers, town boosters, and local clubs backed change because they believed it was their civic responsibility. Typically Progressive, these reformers tempered their idealism with practicality. Most were younger, middle-class, Protestant, and relatively well-educated urban or small-city residents who retained close ties to their rural roots.[12]

Reformers highlighted several causes for the perceived shortcomings of rural life and offered various solutions. Some blamed low commodity profits and advocated greater farm efficiency to boost yields. "The basis of all better rural life," said Great Northern Railroad magnate James J. Hill, "is greater earning capacity of the farmer. By increasing the earning capacity of the small farmers we could readjust their lives and environment, for more comfortable homes, high schools, improved highways, telephones, free delivery by mail, and rural libraries all require money, and cannot be installed or maintained without it."[13] Some critics suggested that rural churches needed to do a better job of confronting the challenges of modern America and work harder to make doctrine more consistent with a "modern" social, and secular, morality of the cities. Rural YMCAs, for example, used religion as a tool to help relieve the shortcomings of rural life. They focused their attention on the physical, social, and educational needs of their communities as well as on spiritual matters.[14] For many, changes in the nation's commodity-marketing system promised greater financial rewards for individual farmers. Others emphasized technology as the best means to free American agriculture from its preindustrial malaise. During the Progressive era, the automobile revolutionized society, and many reformers worked to improve the inadequate rural road systems that hampered the development of consolidated schools, increased transportation costs, and reinforced farm isolation. Groups such as the American Association for Highway Improvement and the Farmers' Good Road League advocated scientific highway management and condemned waste, corruption, and partisan politics that interfered with highway development.[15] Many pointed to the allegedly inadequate system of agricultural credit for investment and advocated new credit systems for farmers.[16]

Whatever their specific emphasis, Progressive agricultural commentators generally agreed that a strong rural economy was the foundation on which to resurrect and restructure agrarian life. Although reformers pursued different paths toward prosperity, most embraced the idea that practical, agrarian adult education provided the key to a better economy and the power to uplift farmers struggling with modernity. Reformers focused considerable attention on the best means of providing educational opportunities that would be an efficient and effective means to train rural Americans to adopt modern economic behaviors.

At first blush, there seemed to be several avenues for delivering adult education to rural Americans. For all the new focus on instruction, agricultural education was not a new phenomenon. Progressives emphasized knowledge as the solution to the problems of declining farm population and the challenges of rural life, but private agricultural clubs and societies had been helping farmers succeed since the late eighteenth century. The Philadelphia Society, for example, was founded in 1785, and the Massachusetts Society for Promoting Agriculture held meetings to improve agriculture and distribute information about new farming techniques as early as 1792. Throughout the nineteenth century, agricultural societies flourished in several states and offered a broad range of educational opportunities. They sponsored fairs, competitions, sales, demonstrations, and classes. They invited lecturers to speak on agricultural topics and published their speeches. During the 1840s, Ohio, for example, created a state board of agriculture that sponsored lectures and courses on scientific agriculture and published pamphlets on farm topics; Columbia University and Rensselaer Polytechnic Institute, in New York state, maintained a pool of experts in agricultural science. In March 1861, innovative local farmers created the DeKalb County Agricultural and Mechanical Society, "the object of which shall be the promotion of agriculture, horticulture, manufacture, mechanical, and household arts."[17]

Perhaps the most important trend in agrarian education emerged at midcentury. In 1854, the Massachusetts State Board of Agriculture concluded that "farmers' institutes" might fill the void in agricultural education. The idea was simple—if farmers could not, or would not, attend classes at a central location, knowledge needed to brought to the farmer. At a farmers' institute, agricultural scientists gave lectures to farmers on a wide range of practical subjects at convenient locations. The state also published pamphlets and booklets covering a range of topics, such as manures, pasture renovation, soils, cattle breeding, grape culture, and butter making. Yale University developed a similar series of courses on agricultural topics that proved so popular that the Connecticut State Board of Agriculture organized its own sessions for farmers. After the Civil War, interest in practical agricultural education increased, and farmers' institutes emerged as the most popular format. During the 1890s, twenty-six states sponsored farmers' institutes on a more or less permanent basis. In 1899, forty-seven states held farmers' institutes with a total attendance of at least 500,000.

The details of farmers' institutes varied from place to place, but they all had common features. Typically, either the state boards of agriculture or the state agricultural colleges and experiment stations managed the institutes. An institute could be held in any location within the state at the request of local farmers—Iowa, for example, required the signatures of fifty local farmers on an institute application. The application was then turned over to a state board of agriculture employee or to a professor of agriculture, who reviewed the request and helped with the arrangements. Expenses for the out-of-town speakers, renting the hall, placing advertisements, and incidental costs were usually paid by those who attended the programs. Some institutes lasted for a week, but more commonly only a couple of days. Daytime sessions typically included lectures and discussions conducted by experts who focused on practical farming techniques, while evening sessions usually consisted of public addresses on topics connected with agriculture. Presentations by innovative local farmers or professionals on unique farming issues in the community sometimes supplemented the lectures offered by the state experts. Lectures were generally short, sometimes only a half-hour long, followed by a question-and-answer discussion period during which farmers could explore issues raised by the speakers—it was usually the case that the discussion part of any presentation was far more valuable to the participants than the lecture portion. Moderators actively discouraged political discussions, as well as "statements based on ignorance, prejudice or superstition."[18] Although practical education dominated the agenda, it was common for institute organizers to include patriotic speeches or recitations and showcase local musical talent. Many institute organizers encouraged women to participate by offering a range of domestic subjects, such as nutrition, cooking, and other family welfare topics. The number of woman lecturers gradually increased as the institutes became more common. New York and Indiana were the first to include topics for farm children. In 1904, North Carolina sponsored institutes for Black farmers. By 1915, farmers' institutes were a regular feature of rural life in many places, with nearly three million people attending more than eight thousand institutes annually.[19]

Illinois was typical of states that supported farmers' institutes. In 1891, the Illinois General Assembly approved a law for the creation of a Farmers' County Institute to assist farmers in holding institutes and develop the agricultural resources of the state generally. The state agreed to pay fifty dollars annually to treasurers of county farmer organizations for the purpose of holding one or more farmers' institutes each year. Institutes were to be at least two days long and to be held in a convenient location within the county. According to the law, institutes "shall be held for the purpose of developing a greater interest in the cultivation of crops, in the breeding and care of domestic animals, in dairy husbandry, in horticulture, in farm drainage, in improved highways and general farm management, through and by means of liberal discussion of these kindred subjects, and in the distribution of papers and proceedings of such institutes."[20] In 1895, Illinois revised its farmers' institute law to establish the Illinois Farmers' Institute, a public corporation, complete with a board of directors empowered to draft bylaws for the government and management of its affairs and three elected delegates from each county farmers' institute organization. The Illinois Farmers' Institute management rules established strict limits on when and how county farmers' institutes could be held. For example, although musical entertainment was an important part of most farmers' institute gatherings, no state money could be spent for such entertainment. Likewise, farmers' institutes could not be held in conjunction with other events such as county fairs, political meetings, or circuses. It is clear that the Illinois Farmers' Institute wanted its programs to be serious affairs, free from partisan distractions, and focused on agrarian education. The Illinois Farmers' Institute proved so popular that the state legislature reaffirmed and amended the statute in 1901, 1903, and again in 1909.[21]

Counties in several states, including DeKalb County, sponsored farmers' institutes that offered farm families short courses in practical farming techniques and home economics, but public support for agricultural education was not limited to farmers' institutes. State and federal departments of agriculture also supported agricultural education. These public bodies published and distributed a wide array of literature about scientific agriculture and the latest technological advancements, sometimes at a minimal cost but usually free of charge. State agricultural colleges and their affiliated experiment stations were especially active in promoting education. They provided an equally broad range of farm bulletins. In many states, colleges also offered a diverse menu of educational services to farmers, including correspondence and reading courses that emphasized the benefits of modern agriculture. Although their emphasis was on increasing production, state colleges were also helping farmers learn new skills—such as business management, economics, and sociology—that promised to make farmers more competitive in the marketplace. At the federal level, the U.S. Department of Agriculture followed the lead of state institutions and developed education programs that mirrored those of the state colleges, although in time its programs would address a wide range of topics, including economics, local community organization, agricultural marketing, and farm tenancy. Local, county, and state fairs highlighted improved agricultural techniques and offered venues for the most successful farmers to show off the practical benefits of their newfound knowledge.[22]

In the private sector, businesses whose success depended on the health of the rural economy—including railroads, mail-order houses, implement companies, banks, and small-town merchants—reinforced the practical advantages of modern agriculture. Some sponsored agricultural shows and prestigious contests that publicized, and sometimes richly rewarded, farmers who raised outstanding crops using the latest farming techniques. Bankers, many of whom believed that inadequate rural education drove farmers off the land, were especially active in lobbying state legislatures to support vocational agricultural courses in local high schools. Farming journals and local newspapers were a regular source of information about innovative agricultural techniques. Metropolitan civic service organizations focused their educational energies on campaigns for the construction and improvement of rural infrastructure, such as schools, roads, bridges, parks, and playgrounds. Women's clubs supported educational activities focused on home economics, health and sanitation, and elementary and secondary education. Countless small towns and villages also formed organizations to address matters of local concern.[23]

Yet, despite the opportunities for education in rural America, many farmers chose not to take advantage of them. Instead of embracing new methods, they often clung to their conservative beliefs and traditional ways. Although many farmers were reluctant to view their farms as business operations instead of homes, much of the blame for the inability to reach out to farmers with helpful advice rests with the educational institutions themselves. Government programs were chronically underfunded and shorthanded. Too often, their scientific bulletins, most of which contained superb information, were overly technical, difficult to understand, or simply inappropriate for local conditions—it is little wonder that few farmers requested them. Journalists wrote extensively on new agricultural methods, but subscriptions to general-interest local newspapers or national agricultural journals alone did not mean that farmers were receptive to the new technological information contained in the articles. Although fair displays and contests winners showed farmers the tangible benefits of modernization, they often did little to explain how success was achieved. Moreover, fairs were important social occasions at which isolated farmers could renew contacts with their neighbors, not necessarily focus on the latest agricultural advice. Contests frequently promoted the products or services of the sponsor as much as general agricultural knowledge. DeKalb County's farmers' institutes were well attended and quite successful, but the quality of institutes in other communities was poor in comparison. Too often they were inadequately staffed, poorly financed, and lacked local support from the commercial sector and were frequently held in locations that were inconvenient to farmers who lived off the beaten path. Many rural folk considered the institutes, like fairs, to be social occasions as much as educational opportunities and were sometimes not prepared to master the technological details that speakers provided. Without the ability of institute speakers to follow-up their autumn or winter institute

lectures with hands-on guidance throughout the growing season, institute organizers found it difficult to influence farmers who remained
skeptical about the benefit of adopting creative agrarian practices.

Progressive-era reformers addressed the inadequacy of rural adult education in several ways. They attempted to improve the farmers' institutes programs through better instructors and improved organization.
They worked to simplify the opaque technical language of the government bulletins and make them more applicable to local conditions.
County and state fairs benefited from better planning and enhanced
emphasis on educational programming. As the first decade of the twentieth century closed, however, it became increasing evident that despite
these improvements, all the educational approaches were limited by a
fundamental problem. They offered rural folk a large dose of complex
material at once, usually during the winter, months ahead of the time
when the farmer could expect to put the new techniques to use. Rural
adult education ultimately relied upon the farmer to take the initiative
to master the material, often during the short time of a farmers' institute
or a county fair visit. Reformers came to realize that agricultural adult
education needed a new model that relieved the farmer's heavy learning
responsibility and was better suited to the needs of the everyday farmer.

One promising technique to make agricultural education more effective was demonstration work, pioneered by Seaman Knapp in Texas in
1902. Knapp had a long history in agriculture as a successful farmer, foreign representative for the U.S. Department of Agriculture, and agriculture educator in New York, Iowa, and Louisiana. In 1902, Secretary of
Agriculture James Wilson, Knapp's close personal friend, asked him to
develop demonstration farms in the American South to educate farmers
about how to manage the cotton boll weevil that was threatening the
economies of several Southern states. The new weevil-control techniques pioneered at Knapp's Porter Community Demonstration Farm
near Terrell, Texas, proved far more effective than traditional methods in
containing the infestation. The success highlighted the value of this
new partnership between professional agronomists and local farmers in
improving farm conditions and preserving the economic well-being of
the community. The program expanded rapidly, fueled by its own success. By 1904, the U.S. Department of Agriculture appointed twenty-four
"special agents" to spread the word about weevil-control methods, and
nearly seven thousand farmers were working as demonstrators. Two years
later, the first official "county agent," W. C. Stallings, was appointed in Smith
County, Texas. By 1908 there were 157 county agents working in eleven

Southern states; in 1910, 450 county agents were undertaking demonstration work in 455 counties in twelve states.[24]

The idea of placing a resident farm expert in local communities was revolutionary. Yet, although the idea was new, it was based on old wisdom. "What a man hears," said Knapp, "he may doubt. What he sees, he may possibly doubt. But what he does for himself, he cannot doubt."[25] Demonstration work proved so successful because, in addition to telling farmers what they ought to do, the expert provided regular and frequent follow-up advising to farmers who experimented with the new techniques. Most agents were well educated, sympathetic, familiar with local agricultural and social conditions, and dedicated to their work. These qualities earned farmers' trust and the support of local civic leaders, which, in turn, made the job of advising farmers easier and more effective. Demonstration was, however, not limited to the fields and barns. Knapp and his colleagues envisioned the county agent as a sort of social engineer who had the skills to implement a fundamental revision of rural life. "[T]here is much instruction [by county agents] along the lines of rural improvement, the better home, its equipment and environment, the country roads, the school at the crossroads, rural society, etc."[26] Home demonstrators worked with farm women to improve their families' health and quality of life. County agents supported "corn clubs" for rural boys and canning and poultry clubs for girls. By 1912, there were 639 county agents working closely with over 100,000 farmers, 67,000 boys, and 24,000 girls. More than any other form of education, demonstration work successfully taught "economy, order, sanitation, patriotism, and a score of other wholesome lessons"[27] to rural Americans struggling to cope with the stress caused by their rapidly changing world.[28]

The success of Knapp's ideas, first in the South and later in other parts of the nation, led Progressive reformers to adopt one of two avenues for the delivery of demonstration-based adult education. At one end of the spectrum, many Progressive reformers believed that the county agent system should become the national standard. Knapp himself even suggested that every county in the nation should have an agent supported by local communities working in conjunction with state land-grant universities and the federal government. In pursuit of that goal, bills were introduced into Congress each year between 1909 and 1914 that were designed to support a national network of county agents. All fell victim to a powerful combination of national political trends of the period, including deep suspicion of federal power, lingering North-South tensions, resistance from jealous federal and state institutions that feared losing power and prestige in the field of adult education, and farmer indifference. Nevertheless, that record of defeat does not mean that farmer education lacked support. Few organized groups were actively opposed to farmer education, and several influential sectors of society were deeply committed to its success. The American Federation of Labor, for example, was very active in promoting agricultural extension. The National Soil Fertility League, an umbrella organization speaking for some 470 separate agricultural associations, chambers of commerce, boards of trade, professional organizations, transportation companies, financial institutions, and manufacturers was also providing support for the idea of a nationwide system of county agents. Private businesses that were affiliated with agriculture—such as railroads, implement dealers, and regional banks—had for years financed agricultural education and were eager to have the government assume that financial burden.[29]

In spring, 1914, Senator Hoke Smith (D-Georgia) and Congressman Asbury F. Lever (D-South Carolina) introduced a bill to create a federal extension service. Controversy over the bill was intense. Smith-Lever supporters claimed that the new law offered something for everyone. Farmers would have better access to information and education that would lead to greater efficiency, higher profits, and better living standards. Agricultural business and industry would enjoy continued prosperity as farm incomes increased. Idealists believed that the law would address their lingering fears that rural America, and the national spirit, was endangered. Support for the bill was strongest in the impoverished South, which had the greatest direct experience with the extension agents.

Outside the South, however, rural politicians were less committed to the law. Some stuck to the notion, not altogether incorrect, that higher yields meant less profit. Others saw the law as a patronizing affront to the sensibility and nobility of their farmer constituents. Some of the bill's critics charged that it was socialistic. A bitter disagreement arose over whether to include black farmers in the program. Many Northern congressmen simply displayed the same indifference as their constituents, who saw little utility in "book farming" and feared that it was yet another burden for taxpayers. Ironically, the bulk of non-Southern support for Smith-Lever came from congressmen who were not primarily interested in farmers at all. Rather, much of the strongest support for rural adult education came from those who saw it as a means of increasing farm efficiency and thereby reducing the cost of living nationally.

Although supply and demand have more influence on the costs of food and fiber than farm production costs, and despite the economic inelasticity of most farm products, urban representatives nevertheless favored Smith-Lever because they were convinced that farmers would pass the lower production costs on to their urban neighbors. Finally, President Woodrow Wilson endorsed the idea and pledged his support. A series of timely compromises moved the Smith-Lever bill through Congress in May 1914. In the end, most in Washington concluded that rural adult education was simply the best means of integrating farmers into the emerging industrial nation. Cost of living issues, not altruism for rural Americans, appears to have been the driving force behind the new law that Wilson correctly described as "one of the most significant and far-reaching measures for the education of adults ever adopted by any government."[30]

The language of the new law reflected many of the lessons that earlier county agent programs had already learned. The law was intended to reach a wide cross-section of society and suggests that extension agents were available to all citizens; but the appropriations scheme indicated a clear congressional intention that the new extension service serve primarily rural people. Congress created a broad mandate that county agents instruct "on subjects relating to agricultural and home economics" through "demonstrations, publications, and otherwise."[31] Although the mission of the nonpolitical extension service was education, it was distinct from other educational institutions, such as state colleges. The extension service granted no degrees, had no fixed curriculum, operated informally with no central campus, and addressed a broad and heterogeneous audience in matters of practical education. Extension agents were encouraged to be flexible and develop programming that addressed local conditions. Participation in extension services was purely voluntary, and the organization had no police power to exercise control over rural participants.[32]

Passage of the Smith-Lever bill was the high tide of Progressive-era farm reform. Modern agriculture, supporters claimed, would be vastly different from its nineteenth-century roots. County agents—working in cooperation with federal, state, and local governments, state universities and land-grant institutions, local clergy, and rural school boards—promised to regenerate American agriculture and restore the nobility of farm life. The traditional countryside—dominated by laissez-faire economics, rugged individualism, inefficient production, and social provincialism—would be made to conform to new standards of economics and organization. Such broad reform, supported by an increasingly powerful and activist federal government, could achieve all the economic advantages of modern agriculture and thereby reduce the costs of food nationally and preserve the country's rural foundation in the modern era. Although Progressive reformers recognized that better organization and efficiency they hoped for would take time, like most Progressive-era reforms it was a remarkably optimistic assumption about the possibility that rural Americans, when led by trained experts, could change their historical social institutions, political practices, and commercial attitudes.

The centralized, national plan for delivering adult education represented by the Smith-Lever Act, however, was not the only model for improving American agriculture. At the other end of the Progressive education reform spectrum, the benefits of education and the practical success of demonstration work led some reformers in a handful of areas to adopt a different approach. Instead of relying on a nationwide system of publicly supported agricultural extension, community leaders developed their own plans for rural adult education. They took matters into their own hands to tailor a program that was suited to their specific needs, rather than hope and wait for a program to be offered to them. While Smith-Lever made extension work far-reaching and national in scope, these earlier county programs were intimate and local. And although the Smith-Lever Act reflected Progressive ideals that stressed centralized, federal responses to social concerns, local reformers in several counties across the nation tapped into other powerful trends of Progressive reform: community empowerment, democracy, and local control of institutions. For these men, the best solutions to problems arose from within the community. Trained experts, working with state or federal experts when necessary but employed by a local, autonomous governing body, provided the ideal way to make farms prosperous and integrate rural society into the broader stream of American life.

The first county to adopt its own program for a locally directed farm advisor was Broome County, New York. On March 20, 1911, the Binghamton Chamber of Commerce hired John H. Barron as a county agricultural advisor, with educational assistance provided by the New York State College of Agriculture at Cornell. Barron, a graduate of the State College of Agriculture, was not the first county agent, but he was the first agent hired by a local consulting organization that was primarily responsible for overseeing the advisor program. In this case, the first "farm bureau" was simply a department of the Binghamton Chamber of Commerce, supported financially by local businessmen, bankers, and the Lackawanna Railroad, with the goal of educating local farmers in "modern" agricultural techniques by a college-trained agricultural expert. The idea of a full-time farm advisor backed by an essentially private agricultural improvement organization sprang up simultaneously in many counties. In March, 1912, the Sedalia Boosters' Club of Pettis County, Missouri, hired Sam M. Jordan as the county advisor with funds provided by the local county court, the Sedalia school board, local subscriptions, and Sears, Roebuck and Company chief executive and philanthropist Julius Rosenwald. Pettis County was also the first county to create a completely separate governing body, the Bureau of Agriculture, to oversee the new farm advisor. The Pettis County Bureau of Agriculture was not a general membership organization, but rather a group of representatives from each of the county townships, six officers, and representatives from the local school districts. It also had three "honor" committees: a farmers' committee of men whose farming practices preserved soil fertility, called "the Soil Builders"; a "Good Stockman" committee of farmers who raised purebred stock; and the "Road Builders" committee open to farmers who helped maintain the county roads. In April, 1912, Chemung and Jefferson counties, New York, both had private associations that hired and oversaw farm advisors. In the summer of 1912, Cape Girardeau County, Missouri, hired its own county advisor, and the Better Farming Association of Bottineau County, North Dakota, hired M. B. Johnson as its advisor. Counties in Vermont, New Jersey, Minnesota, Iowa, West Virginia, and Indiana all began the process of hiring farm advisors in late 1912 and early 1913.[33]

At first, these new farm advisors were usually hired and paid by chambers of commerce, banks, national companies, and other commercial organizations. Farmers, in fact, had little voice in determining policies or shaping programs. West Virginia and New York were the first states to recognize the importance of farmers to the success of their advisor programs and arrange for them to become real participants in the planning process. Typically, before an agent was hired in a county, local farmers and other boosters pledged financial support, usually a one-dollar membership in a "county farm bureau." The bureau elected officers, recruited new members, assisted the advisor in developing and implementing programs, and directly and indirectly supported the farm advisor in a variety of unique ways. Counties that had begun their farm advisor programs without farmer support quickly found that they needed to amend their procedures to include them. Farmers and other advisor supporters in Broome County, for example, organized the

Farm Improvement Association of Broome County in 1913, which assumed the responsibility for managing the county's farm advisor program—a year later the Farm Improvement Association became the Broome County Farm Bureau. This new arrangement enabled farmers to have a voice, along with the commercial interests in the county, about their own local affairs. Perhaps more important, however, it gave farmers a stake in the farm advisor experiment that made them more receptive to the new techniques farm advisors introduced.[34]

Another one of the counties, and the first in Illinois, to organize a general membership agricultural improvement group, enlist farmer support, raise funds, hire a farm advisor, and adopt its own program of comprehensive county agricultural education and service during this formative period was DeKalb County, Illinois.

2 | THE "SOIL IMPROVERS"

• In the years prior to World War I, Progressive-era reformers waged a far-reaching reform campaign against what they believed to be the inadequacies of rural life. They saw the agricultural sector rooted in a lethargic, preindustrial, and provincial past that threatened to derail the nation's dynamic embrace of modernity. For Progressives, the problem and its solution seemed clear. Reformers relied on modern standards of economic efficiency, social organization, professionalism, and government activism to remedy the appalling conditions they found down on the farm. Their plan envisioned new financial practices to enhance farm efficiency and revolutionary changes in rural social institutions. Improved production, in turn, would boost farm incomes and elevate living standards for rural folks while lowering the costs for food and fiber for all Americans. The key to this plan was agricultural education. To integrate rural America successfully into the rapidly evolving industrial economy of the time, farmers would need to be taught new, scientific, agricultural methods, and reformers assumed that rural folk would embrace modern agriculture once they understood its economic and social benefits. At the national level, that reforming impulse resulted in the Smith-Lever Act of 1914, which created a nationwide rural educational infrastructure, supported by the federal government, that coordinated state and community institutions into a comprehensive educational network led by a trained expert, the county extension agent. In DeKalb County, Illinois, community leaders and many area farmers also embraced the Progressive education spirit. Unlike the other Progressive reformers, however, DeKalb residents did not look to Smith-Lever and the federal government for assistance to improve their farms. Instead, they took an independent path toward agricultural education.

Residents of DeKalb County valued agricultural education long before it became a core tenant of Progressivism. Throughout the second half of the nineteenth century, DeKalb-area farmers attended agricultural education programs designed to help them succeed. The first county fair, called the Union Agricultural Institute, was held in Sandwich, in 1858. The DeKalb County Agricultural and Mechanical Society flourished during the 1870s. In March, 1890, fifteen Afton, Illinois, farmers organized the "Afton Center [Farmers'] Alliance" and became part of the "grand army moving forward toward victorious reform." Newspapers ran articles highlighting exceptional farmers. The *Sycamore Tribune*, for example, ran a feature in 1904 about the John Anderson farm, located south of Sycamore, that was the "best forty acre field in the area," producing almost ninety bushels of corn per acre.

During the early years of the twentieth century, however, farmers' institutes and college short courses were the most comprehensive way to reach out to DeKalb County residents with information about new farming techniques. The first recorded DeKalb County Farmers' Institute was held in Genoa, in 1903, featuring speakers on farm topics, new farm products, and "ladies' fancy needlework." A year later Somonauk hosted an institute that included lectures and discussions, exhibits of corn and other farm products, prizes for the best crops and stock, and a speech titled "Legislation for Farmers." Local newspapers urged farmers to attend the institutes: "Farmers are apt to complain that nothing is ever done to help them through governmental agencies, yet here is an agency of the utmost practical benefit, created wholly for them, which they largely neglect to make use of." Institutes provided educational opportunities in the county at least annually thereafter. In 1907, institute organizers held three institutes in three towns. A year later, six different county communities hosted farmers' institutes. The Northern Illinois State Normal School in DeKalb offered the first in its series of annual short courses on agriculture in 1909. The next year, 1910, farmers' institute organizers arranged for six one-day institutes in different towns. In 1911, the farmers' institute, under the leadership of its president George Gurler and secretary Henry Parke, set an ambitious schedule of eleven institutes in different communities between December 5 and 16. The institutes, entitled "Better Than Ever Before," were intended simply, as Gurler put it, "to fill the farmers of this county as full of agricultural knowledge as they will hold."[1]

The 1911 the farmers' institutes addressed a wide range of issues related to agricultural production and the improvement of farm life. Topics for the sessions included crop rotation, bridge and hard-surface road construction, techniques for "building up a run down farm," home economics, "good air and good health," "domestic science clubs" for rural women, and "diet for children." Speakers also presented lectures on soil depletion and renewal. This was a topic that was particularly important, because farmers and community leaders alike recognized that preserving soil quality was the key to ensuring the county's prosperity. Locals hailed the institutes as a rousing success. The *DeKalb Daily Chronicle* noted that there "was a large crowd of interested farmers" at the Kirkland Institute, and in Sycamore there was "a monster session being held which is being attended by scores and scores of farmers."[2] With typical hyperbole, it reported that the institute held in Elva, on December 12, was the "most successful one-day institute ever held in this county or any place else. . . . The attendance was remarkable, the program was simply great, the competition for the various exhibit prizes was keen and the enthusiasm in the movement unbounded."[3] The farmers' institute momentum was at high tide in the county, and residents were clearly proud of their programs. "DeKalb County," boasted a local newspaper, "is a pioneer in the most approved methods of conveying information and instruction to the farmers by means of short term schools in agriculture, and the splendid efforts of the men who have affairs of that nature in charge in this county have resulted in wonders as regards the development of the farmer's institute idea in the state."[4]

The success of the December farmers' institutes was due to not only the popularity of the course offerings, but also the support the programs received from several local farm and municipal organizations. Some of the most important grassroots backing came from township farm clubs in the county. The first of these clubs, the Sycamore Alfalfa Club, later the Sycamore Farmers' Club, was incorporated in 1909.[5] By the end of 1910, DeKalb County had seven township farmers' clubs, with more than 600 farmer, merchant, and banker members. These community clubs, formed from the "ground up" by interested local citizens, served several important agricultural and social functions, including sponsoring local fairs. Clubs also provided educational materials and information to their members, established community centers for the discussion of local issues, and generally served to connect farmers with other county organizations. One of their most important roles at the time, however, was their close relationship with the DeKalb County Farmers' Institute. Each club, for example, provided a representative to the institute and served as a talent pool of local speakers to supplement the outside experts who spoke at the farmers' institutes. Beyond the tangible benefits of improved social, moral and educational standards, the clubs offered farmers a sense of pride in their work. "One finds fewer weeds in the fence corners," said Parke in 1911, "cleaner lawns, neatly dressed families, more pride in every way; a better social spirit exists; distances between farms have been blotted out by a more neighborly feeling." The cooperative spirit among club members also resulted in more farming and marketing collaboration, and even, some believed, improved political standards.[6]

Considerable financial and business support for farmers' institutes came from area businessmen, especially bankers. Practically everyone in the county was personally connected, in one way or another, to a family farm. As landholding commercial institutions, banks, especially, shared with their rural neighbors a clear understanding that their financial success depended on the continued prosperity of area farms. Many bankers were also part-time farmers or personally owned farms. According to the Illinois Bankers' Association, in 1911 bankers in Illinois owned or directly represented more than one million acres of farmland. Often, their support for rural adult education came via informal channels. Bankers, for example, had always been well represented in farm education—two steadfast supporters of the DeKalb County Farmers' Institute educational outreach programs were Arthur Dodge, from Malta, and Dillon S. Brown, from Genoa, both of whom were influential county bankers and successful farmers.

DeKalb County bankers were also members of a larger statewide banking association, called the "banker-farmers," that was an informal subgroup of the Illinois Bankers' Association composed of rural bankers. These men shared a common interest in agriculture and were impressed by the success of Seaman Knapp's demonstration work in the South. At the Illinois Bankers' Association meeting in Springfield, Illinois, in September 1911, the banker-farmers heard from the Great Northern Railroad's James J. Hill about the success of Knapp's county agent program. "The soil tiller is as numerous, as much in need of instruction, and as unable to leave home in search of it as a child. The education," said Hill, "must be taken to him." The Illinois bankers who had invited Hill were as deeply committed to demonstration work as was their keynote speaker. They believed that the existing methods for delivering agricultural adult education—especially the farmers' institutes and the outreach programs sponsored by the state agricultural colleges, universities, and experiment stations—were "only superficial and frequently fruitless" because only a small percentage of farmers learned the new techniques, and still fewer applied them. "The methods of greater earning capacity of the farmer and soil have all been discovered at great cost," noted Hill, "and can now be easily taken directly to the farmer, at small expense, by the Field Demonstration Plan." Demonstration advocates highlighted Knapp's success in eradicating the boll weevil infestation and the fact that much of that success was due to the county agent, who kept close track of demonstration farms and encouraged farmers to see for themselves the benefits of new agricultural techniques. The DeKalb County bankers also endorsed the idea of local demonstration and experimental plots where farmers could see for themselves the tangible results of new scientific and technological advances in agriculture. "We cannot blame the farmer for wanting to be 'shown,'" wrote a sympathetic county reporter, "and we do feel that such demonstration work helps to show him and his neighbors, as no bulletin or other system possibly can do."[7]

In 1911, the Illinois Bankers' Association, which was originally organized into ten regional groups, reconfigured itself into county subdivisions, a move "designed to bring about a closer relation between local banking institutions." As part of that restructuring, twelve DeKalb County banks organized into the DeKalb County Bankers' Association in early October 1911 to better improve the professional relationships among county banks and coordinate their community service efforts. They would also actively recruit the remaining nine area banks into the association in the near future. These county bankers were especially concerned about the prospects of the county's slowly declining soil fertility. Improved farming techniques and management, they believed, would halt depletion and restore fertility, improve farm efficiency, and produce better yields at lower cost. Soil exhaustion also impaired the value of rural real estate that most area farmers pledged as security for their loans.[8]

The last group to lend its support to the farmers' institute programs included local journalists and their newspapers. Their uniformly favorable stories about the farmers' institutes reported the events and advertised the programs. They also recognized that regional farm prosperity translated into wider distribution, higher subscription rates, and increased advertising revenue. In October 1911, editors of all but five of the county newspapers joined together, as the bankers had done, to form the DeKalb County Newspapermen's Association to better coordinate their journalistic efforts and promote community causes, especially the farmers' institutes.[9]

The activities of these three organizations—the local farm clubs, the bankers, and the journalists—to promote agricultural education reflected a typically Progressive pattern of reform. Like many reform movements of the era, leadership for rural education came from powerful and influential elites who banded together in professional organizations to

lend their collective voice in support of change they deemed was necessary for the good of the whole community. In DeKalb County, leadership came primarily from successful farmers and business leaders who
shared a common interest in the economic well-being of their community. Moreover, in turn-of-the-century agricultural America, including
DeKalb County, community elites were often deeply involved in both
farming and commerce. Although each organization represented different constituents, they were united in their common interest in agrarian
education. Instruction, they believed, should be practical, timely, and,
above all, accessible to farmers so they could boost their profit margins,
sustain the integrity of the soil, improve livestock, and, in general, celebrate rural living. Their goal was not to recreate a misty agrarian vision
of the past, but to move forward boldly and embrace the modern world
and all the scientific and technological advances it offered. Only by doing so could the county's prosperity be assured. Indeed, some publicly
minded men, like Dodge and Brown, moved easily in all three groups.
Some served, as did banker J. B. Castle of Sandwich, in the governing
structure of the various clubs and associations. Not only did these community leaders and their organizations offer substantial tangible and in-
tangible support for the farmers' institutes, they also provided the vision
and drive to extend Progressive agrarian educational ideals in a new direction in DeKalb County.[10]

Farmers' institutes in DeKalb County always enjoyed considerable
support. Yet, despite the general success of the institutes and the enthusiasm with which area residents embraced the programs, it became increasingly evident to many county leaders that the farmers' institutes
alone were not sufficient to deliver the kind of new agricultural education necessary to preserve the region's prosperity. Institutes were held at
least annually, but often farmers considered them as much social occasions as opportunities to learn. Frequently, the fact that institutes were
held in the late autumn or winter, when farm work was less hectic,
meant that there were several months between the lectures and the time
when the new techniques could be put to use. The greatest drawback,
however, was that the intensive, lecture-oriented programs provided
farmers with new theories but little practical direction about how to apply that knowledge in the field. It was clear that successful continuing
education required more than once-a-year farmers' institute instruction.

These problems caused many community leaders to consider new
ways to make agricultural advice available on a regular basis. The most
popular idea during 1911, especially as the year went on, was to hire a

permanent farm advisor along the lines of the county agent program pioneered by Knapp and advocated most persuasively by members of the banker-farmer movement. The proposal was championed in its earliest stages in DeKalb County by several influential men, most notably Charles Bradt, a DeKalb banker, Fred Townsend, a banker in Sycamore, and most enthusiastically by banker and farmer Brown and Parke, a college-trained biologist and successful farmer. Their decade-long association with the farmers' institutes and close ties to both the farm and business communities led them to conclude that the county needed to hire an advisor who would be an expert on agricultural conditions in DeKalb County and devote his time to delivering the latest information and expertise to area farmers. By late autumn of 1911, the pursuit of an advisor had become a crusade in DeKalb County. Several bankers and newspapermen supported the idea of hiring a full-time farm advisor and were already refining and promoting their proposal. The flurry of organization among the bankers and journalists in 1911, coming as it did in on the heels of the successful farm clubs, was not a coincidence. Although each professional association had several motives for organizing, their most immediate focus of attention was to raise local interest in the new farm advisor. Their early promotional efforts showed promising results, and by late 1911 there was considerable interest among a broad spectrum of community leaders and many area farmers in a permanent farm advisor.

One of the most important issues that arose during the planning stages for an advisor was just what sort of agricultural expert should come to DeKalb County. Despite the strong personalities and diverse financial interests of those most deeply involved in the project, there was surprisingly little disagreement. It seemed clear to everyone that soil exhaustion was the most consistent and potentially devastating long-term problem confronting DeKalb County farmers. Seven decades of cultivation had resulted in serious soil depletion through erosion and nutrient loss—the fantastic fertility of the virgin prairie soil that had so awed pioneer farmers was long gone. The evidence was obvious; farm fields no longer produced at the rates that had once astounded earlier settlers. The increasing mechanization and commercialization of farming and the need to achieve maximum yields accelerated the pace of decline throughout the late nineteenth century. Traditional methods of fertilizing with manure and other natural materials helped, but many in DeKalb County feared that the soils could not long sustain agricultural production at a level necessary to ensure the county's continued prosperity in the modern era. Without proper soil-management techniques, long-term prospects for the future appeared gloomy. Consequently, all agreed that the first duty of a farm advisor would be to help farmers restore and preserve the soil.

Influential supporters of the advisor idea also believed that their county agent would need to be supervised by some kind of autonomous governing body, but they were uncertain about whether an existing organization would hire and oversee the new farm advisor or a new governing structure should be created for the task. On December 10, 1911, Parke sent letters to the directors of the DeKalb County Farmers' Institute, the DeKalb County Bankers' Association, the DeKalb County Newspapermen's Association, and local farm club officers, inviting them to attend an organizational meeting on January 5, 1912, to discuss details of the farm advisor proposal. Two weeks later, on Christmas Day, representatives from the farmers' institute and the two professional associations met in Sycamore to draft an agenda for the January meeting. By then, it was clear that most supporters wanted a new, independent organization to govern the farm advisor program. The purpose of the January meeting, they said, was "to organize a soil fertility league" in the county that would "hire an expert along soil lines to be in the employ of the farmers of this county exclusively and to analyze soils that seem to need some kind of treatment, tell the farmers what their land needs and to give any other advice that may be necessary in that line." Their economic motivations underlying the plan were equally clear. If their proposal were successful, "the farmers of this county would be able to add hundreds of thousands of dollars of value to their lands by increasing productivity."[11] Yet, while economics was the foundation for the farm advisor movement, from the beginning supporters of the farm advisor ideal believed that the services of the new farm expert should be available to any farmer in DeKalb County, whether land owner or tenant, association member or not.

Although leaders of farmers' clubs and area businessmen were united in their desire for a full-time farm advisor, they also recognized that without broad community support their plans would probably fail. To encourage public participation, they issued an open invitation to DeKalb County residents to attend the January 5 meeting. They hoped, of course, for strong farmer turnout, especially from the southern part of the county where the idea of a permanent farm advisor was met with greater skepticism. During the ten days between Christmas Day and the January meeting, journalists aggressively promoted the farm advisor idea and the organizing meeting through articles and editorials. Two days before the meeting, the *DeKalb Daily Chronicle* headline revealed how much advance planning and decision making had already taken place. The soil expert, it reported, will have "his headquarters in some central point, probably DeKalb, where he can have access to a good chemical laboratory, and to be at the service of the farmers of the community who are confronted with puzzling problems as to why their farms are not growing as much corn as they ought to per acre." To press home its support, the *Chronicle* indulged in the sort of enthusiastic boosterism that characterized so much of the local reporting on the subject. "This man could analyze soils, tell what elements they need and advise the farmer what to put on them with the result that this county which is now well up on the list in Illinois would lead in soil fertility as it is going to in everything else."[12] A day before the meeting, the paper announced on its front page that farmers and other influential men from all over the county were gathering to discuss a farm advisor who would run "a county bureau which will work for the interests of the farmers of DeKalb County entirely."[13] The paper also advertised the keynote speakers invited by the sponsors to boost attendance at the meeting. The first notable was the "broad-minded financier" B. F. Harris, president of the Illinois Bankers' Association, who had long advocated scientific agricultural methods to boost farm incomes. O. D. Center, state superintendent of farmers' institutes, was also scheduled to speak. The third guest was one already familiar to many DeKalb County residents, University of Illinois professor William G. Eckhardt, who had just spent ten days lecturing DeKalb County farmers about soil fertility at the December 1911 farmers' institutes. They were all expected to endorse heartily the call for a "soil improvement association" and a full-time farm advisor.[14]

Community leaders were so interested in generating widespread community support because they understood that a full-time farm advisor was an expensive proposition, and they needed sufficient funds to fulfill their vision. So, while advocates generated public interest in a farm advisor, they also worked diligently, and often quietly, to secure the money that would be necessary to hire a soils expert and furnish his office and laboratory. They reckoned that day-to-day expenses for an advisor could eventually be raised through some kind of membership dues in a private organization, or even through public donations, but the cost estimate for start-up and farm advisor operation for the first year was at least $10,000. For that, organizers needed an immediate infusion of cash until the rest of the financial details could mature. Otherwise, the project was doomed. As a banker and a chief spokesman for the project, Brown was in an ideal position to solicit financial support from his fellow bankers in the county. It would likely be an easy job, because most of the bankers were already members of the county bankers' association, involved in the statewide banker-farmer movement, or at least endorsed its goals, and had been committed to a farm advisor for DeKalb County for several months. Now, however, was the time for them to make good on their commitments. On the brutally cold New Year's morning of 1912, Brown and Frank Olmstead, a local merchant, set out in Olmstead's roadster to call on as many county bankers as possible. Although the thermometer read only ten degrees below zero that frosty day, Brown and Olmstead drove 120 miles across the county and called on every banker except two. Most important, they convinced the twenty bankers they met to commit $100 each year for three years to support the farm advisor.[15]

January 5, 1912, the day of the organization meeting, dawned even colder, but people from all over the county braved temperatures that were "cold enough to freeze the smoke of a pipe" and gathered in the Elks Club, upstairs from the *DeKalb Daily Chronicle,* to attend the organizing meeting that would "result in a work of paramount importance for the people of this part of the country." The meeting was part business meeting, part farmers' institute, and part pep rally. Keynote speaker Harris stressed his familiar theme of preserving soil fertility. "The great State of Illinois has grown rapidly rich, and woefully wasteful," he thundered to the crowd, "and we have been spoiled with an over-rich soil, which is now crying for better treatment, improved methods that will halt its declining fertility and bring greater yield, at less cost, for every bushel added to the yield is almost clear profit." The next speaker, Center, provided a more technical analysis of the economic benefits of improved crop yields and efficient stock management. Eckhardt, the last outside speaker, reminded his listeners of the value of practical farm advising and highlighted the success of demonstration and experimental farms across the nation. Locals Parke and Brown also took the floor and voiced their support for a farm advisor.[16]

By afternoon, the crowd was ready to get down to the practical work of organizing and fund-raising. On a unanimous vote, the crowd agreed to a plan, drafted earlier in December, by which an interim executive committee would manage the fund-raising and oversee the appointment of a farm advisor.[17] The crowd also approved a subscription system to support the advisor and appointed three men from each township to approach farmers in their areas and solicit contributions. Under the plan, the recommended contribution rate for the advisor was $5 for each quarter section (160 acres) of land. In addition to the $2,000 already raised from bankers by Brown and Olmstead a few days earlier, DeKalb Township pledged $500. With the organization in place and funds beginning to flow in, all that remained was to find someone for the job.[18]

The ad hoc organization plan approved in early January was adequate to address immediate issues, but it was clearly not a solid foundation on which to build a comprehensive county advisor program. Consequently, on January 20, representatives of the farmers' institutes, the bankers' association, and the newspapermen's association met again to finalize the details of their new organization. According to the minutes of their meeting, "Moved by D. S. Brown, seconded by Bradt that the combined associations of the DeKalb County Farmers' Institute, the DeKalb County Newspapermen's Association and the DeKalb County Bankers' Association be incorporated as the DeKalb County Soil Improvement Association."[19]

This new organizational structure was better equipped to meet the problems that emerged as the complex details for the farm advisor were finalized, and it moved quickly to hire an expert who would be available to advise farmers during the upcoming farming season. It appointed a delegation consisting of Bradt, Parke, and Townsend, all prominent members of the DeKalb County community, to meet with Eugene Davenport, dean of the College of Agriculture at the University of Illinois, and solicit his suggestions for suitable job candidates. Davenport pledged the full support of the university and immediately nominated William G. Eckhardt for the farm advisor position. Eckhardt was well-known in the Illinois Farmers' Institute organization and had an impressive resume—he had been educated in crop and soil management, was a protégé of the famous soil scientist Cyril G. Hopkins and an experienced teacher, and operated twenty agricultural experiment stations throughout the state. It was an added bonus that Eckhardt, a regular lecturer at DeKalb County Farmers' Institute, had spoken at the Soil Improvement Association organizational meeting the previous January and was already familiar with the DeKalb area and its residents. Davenport told the delegation that he would urge Eckhardt to accept the position, but not without adequate, long-term funds to support the work. The DeKalb delegates assured Davenport of their financial commitments and offered Eckhardt the job. After three weeks of deliberating, Eckhardt turned them down, despite the fact that the $4,000 salary offer was more than double what he earned as a university teacher. He loved his university career and teaching. He was also concerned about accepting a job for which there was no precedent. Nevertheless, on February 9, 1912, under gentle but persistent pressure from his university supervisors and the DeKalb representatives, Eckhardt finally accepted the job, but only after his academic advisors Davenport and Hopkins and University of Illinois president Edmund J. James agreed to serve as an advisory committee and support this new and challenging form of agricultural education. They also granted Eckhardt a three-year leave of absence from his university position, in case things in DeKalb did not work out. Eckhardt's starting date was set as June 1, 1912, at the close of the university's academic year.[20]

Once the "Soil Improvers" hired Eckhardt, the Association confronted several difficulties in the spring of 1912. One challenge was to agree on the location for the advisor's office and laboratory. The Soil Improvement Association members from Sycamore, the county seat, believed that the advisor should be based at the new county courthouse in their town. They had powerful support from such men as Brown, Townsend, and Parke. Others, perhaps hoping to boost the prestige of DeKalb and uneasy about the fact that Sycamore is located in the northeast part of the county, believed that the advisor should have his offices in DeKalb, which, though still in the northern part of the county, was closer to the geographical middle of the county. Such a more central location in the county's largest city would make it easier for the advisor to reach all parts of the county. It is also likely that some DeKalb supporters believed that locating the advisor closer to the middle of the county might increase public enthusiasm for the plan in the southern townships, where the farm advisor program was viewed with some skepticism. All the contestants, however, were anxious not to let their differences compromise their common goal, and the dispute was settled "harmoniously" in favor of the more central DeKalb location for the advisor's office. As a compromise, the Soil Improvement Association would be headquartered in Sycamore.[21]

Another difficulty that arose, however, could not be so easily resolved. In late March, the DeKalb County Board of Supervisors agreed to place the county's farm, sometimes referred to as the "poor farm," at the disposal of the new advisor to use as a demonstration farm. In April, they voted to hire Eckhardt "to supervise, manage and control the entire farm operations of the County Farm of DeKalb County" for three years at an annual salary of $2,000. The new position would cover a significant part of his salary expenses. The benefits of DeKalb County government involvement in the farm advisor project were obvious, but the offer also raised a troubling problem. The original intention of the county advisor advocates was for the advisor to be independent of political control and to provide the best agricultural advice free from outside interference. The board of supervisors' offer, however, raised the prospect that Eckhardt's position might become politicized. As a salaried county employee, there was a possibility that the supposedly independent county advisor would be answerable to the county board of supervisors and subject to their interests and goodwill. Although the action by the supervisors was a well-intentioned offer designed to help support the new advisor financially and provide a location for his demonstration plots, the prospect that their new soil expert might be answerable, directly or indirectly, to groups other than the Soil Improvement Association, which might have conflicting political or economic interests, concerned Eckhardt and his supporters. In the end, the offer, and money, proved too tempting to refuse, but the uncertain relationship between the private Soil Improvement Association and the elected board of supervisors remained. By contrast, the Soil Improvement Association was far more certain about offers of financial assistance from outside the community. It did not hesitate to reject a $1,000 offer from the Sears, Roebuck and Company chief executive and philanthropist Julius Rosenwald, for example, to preserve its cherished sense of local self-control.[22]

The most pressing problem of all, however, was fund-raising. With Eckhardt's acceptance of the job contingent on procuring the necessary money, securing commitments for at least three years was absolutely necessary. After that, supporters believed, their new advisor would be so deeply ingrained in the DeKalb agricultural community that his continued support would be assured. Despite the valiant efforts of organizers throughout the winter and early spring of 1912, fund-raising met with less success than supporters had hoped for. Although many people were happy about the prospect of a farm advisor in theory, not everyone was equally interested in contributing money for the cause. Advisor advocates canvassed personally and enlisted the DeKalb Newspapermen's Association to redouble their efforts to drum up support. Parke was especially active in campaigning for the farm advisor. On February 9, 1912, Parke met with county editors to brief them on the Soil Improvement Association and exhort them to use their influence to build support for the advisor among their readers. In his speech to the journalists, Parke told them that they should share in the responsibility for its success and outlined his vision based on collective action. "[W]e have a great cause to work for," Parke told the writers, "that of making DeKalb County the banner county in the state. Let us all pull together—the editor, the banker and the farmer and secure for DeKalb County that prize, which surely will reward us better farmers, larger yields, higher land values, better roads, better schools, better industrial conditions and a higher standard [of living]."[23]

Parke also lobbied local ministers to "inject in your sermons thoughts that will stimulate better and more scientific methods of farming" and reminded them that "sermons on corn have saved many souls." "Will you lend your personal aid in the pulpit and in your parish in this great movement for DeKalb County, which has for its object the building up and maintaining of our agricultural resources[?]" As he did with the newspapermen, Parke told the clergy that collective action by all the leading sectors of the county, including its religious leaders, to improve farming in the area would naturally lead to improved living standards and elevated morals for all. Parke noted that the movement would result in "better farms, better farmers, better churches, better schools, better citizens and better industrial conditions." "With our four organizations, agricultural, financial, editorial, and religious riding in one band wagon, what could we not accomplish? We can make DeKalb County known around the world." On a more somber note, Parke also reminded listeners about the consequences of not supporting the new Soil Improvement Association. "Your bread and butter," he said, "comes from the farmer. He feeds the world, the laborer, the merchant and the millionaire. When he fails, we all fail."[24]

Despite such passionate pleas, funding the county agent program remained a greater problem than was reported by the upbeat press or in the public meetings. After the bankers and the county government had pledged their support—totaling $4,000—the local government for DeKalb township pledged $500 a year, and South Grove township government pledged $300. Other townships slowly but consistently raised money. The balance was divided equally among the farmers in the county's nineteen townships. A three-member committee in each township asked farmers and landowners for their support. Some of the fund-raising committees, however, found it difficult to convince township governments to support the Association directly or persuade skeptical farmers to contribute. Financial support for the program was weakest in the less-populated southern part of the county, and supporters continued to concentrate much of their promotional efforts there. By the end of May 1912, about 700 area farmers, around one-quarter of the farmers in the county, had joined the Association, but it was still more than $2,000 short of its $10,000 goal. It also was in debt to the Citizens' National Bank for a $400 loan it had taken to provide cash for immediate organizing needs.[25]

Mr. Hammond's late-model Marsh harvester hitched to Charlie Broughton's Buick at the Whitmore-Oakland farm, between Sycamore and DeKalb, Illinois, July 8, 1911. The car "walked away with the harvester as if it were nothing."

right—"Old-timers" demonstrating manual harvesting techniques with hand-cradle scythes and rakes at the Whitmore-Oakland farm, July 8, 1911.

below—DeKalb County Soil Improvement Association certificate of membership for H. A. Lanan, June, 1912. The document is signed by Dillon S. Brown, Henry Parke, and William G. Eckhardt.

Eckhardt, with his roadster, offering advice at the Nichols farm, 1912. Left to right: Clarence, Ruth, Ira, Charles, and Howard Nichols, Eckhardt. Eckhardt is holding a soil auger that he used to take soil samples; the box on the running board probably contained the chemicals and litmus papers he used to test the soil. Note the chains on the rear wheels of the roadster for traction on the muddy farm roads.

William G. Eckhardt, first farm advisor for the DeKalb County Soil Improvement Association, c. 1912. Eckhardt began his duties as the farm advisor on June 1, 1912, and continued in the position until he resigned in April 1920.

Lincoln Watson, of Afton Township, and his team with a fifteen-bushel load of "pure" red clover, alsike clover, and alfalfa seed in front of the DeKalb County Soil Improvement Association's first office, on North Third Street, DeKalb, March, 1914. This was Eckhardt's office. The Association's main office was in the Pierce Trust and Savings Bank of Sycamore, where it held board meetings until 1920. The Soil Improvement Association sold red clover seed in the spring of 1914 for $7.80 per bushel, about $2.70 less than comparable seed on the open market.

Henry H. Parke speaking to a group of DeKalb County farmers attending the third annual DeKalb County Farmers' Picnic at the DeKalb County Farm, September 12, 1914. The banner around the base of the podium reads, "The wealth of Illinois is in her soils, and her strength lies in its intelligent development."

The DeKalb County Farm Committee at the DeKalb County Farm, standing in the year's fourth crop of alfalfa, 1914. Left to right: John Latimer, Frederick B. Townsend, and Chris Awe. Latimer and Awe were farmers and members of the DeKalb County Board of Supervisors. Townsend, a banker, was a member of the executive committee that created the new DeKalb County Soil Improvement Association and was also a member of the subcommittee—along with Henry Parke and Charles Bradt—that hired Eckhardt in the spring of 1912.

Edna Minard Baird, Eckhardt's secretary, in the DeKalb County Soil Improvement Association office, 1914. On the wall above Baird's desk are samples of corn and alfalfa Eckhardt used to help instruct farmers about modern farming techniques. Abraham Lincoln appears on the Anderson Brothers Clothiers and Furnishers' advertising calendar.

Henry H. Parke with soybeans, c. 1915. Parke, "father of the Farm Bureau," was a leading figure in the farmers' institute movement in DeKalb County and the most influential and persistent spokesman for the Soil Improvement Association. He served as the secretary-treasurer of the Association from its creation in the spring of 1912 until January 1922; vice president between January 1922 and February 1923; president from February 1923 until December 1924; and vice president from December 1924 until April 1926.

right—Dillon S. Brown, a veterinarian and banker from Genoa, Illinois, and one of the leading supporters of a permanent farm advisor. He helped arrange the Soil Improvement Association's start-up funding and served as the first president of the Association, from May 1912 until February 1923.

below—Farm advisor Eckhardt demonstrating the difference between corn grown in improved soil (left) and unimproved soil (right), c. 1915.

Dillon S. Brown (sixth from left) with young associates in front of the DeKalb County Soil Improvement Association's North Third Street, DeKalb, office, c. 1915.

Alfalfa demonstration at a farm near Waterman, Illinois, September 27, 1915. Assistant U.S. Secretary of Agriculture Carl Vrooman attended the event, along with other local and state dignitaries.

The Soil Improvement Association Decennial Celebration, June 29–30, 1922. Altgeld Hall at the Northern Illinois State Normal School is in the background. Between 20,000 and 30,000 visitors celebrated the Soil Improvement Association's anniversary and the farm bureau movement with speeches, bands and choirs, floats from fifty-five midwestern counties (DeKalb County's float theme was "Pure Seed for DeKalb County"), and a four-act play written for the event.

U.S. Secretary of Agriculture Henry C. Wallace speaking about principles of the McNary-Haugen farm relief bill at the annual Soil Improvement Association picnic at "Anne's (Glidden) Woods," August 31, 1923. Wallace's private conservation with farm advisor Tom Roberts and Robert's associate, Charles Gunn, convinced the pair to begin experimenting with hybrid corn.

Threshing on a farm near Malta, Illinois, August 4, 1925. DeKalb-area farmers began switching from draft animals to tractors as early as 1908. Note the long, exposed, and hazardous drive belt from the tractor to the machine.

A "big team hitch" turning over a field of clover with a plow and harrow at the same time, c. 1930s. Clover provided nitrogen to improve soils after decades of farming.

Horse-drawn spreader applying limestone to a field, c. 1930s. In the early days, farmers applied limestone to fields by hand with shovels. One of the most common soil prescriptions for DeKalb County fields was limestone. A typical Eckhardt soil prescription for depleted fields read, "Limestone, 2 tons per acre anytime this fall or winter. Follow in the spring with 14 pounds of clover seed per acre. When you get the soil sweetened and the clover to supply the nitrogen, you won't be satisfied with anything less than 75 or 80 bushels of corn to the acre every year. By that time, you will be ready to put on some rock phosphate, and after that 100 bushels won't be too much to expect."

As late winter gave way to spring 1912, DeKalb's experiment was being closely watched and had already attracted considerable attention, both within the state and across the Midwest. The *Farmer's Voice*, for example, reported DeKalb's progress toward a full-time soil expert in late February. In April, Brown gave the keynote address to the Elgin Commercial Club, located in neighboring Kane County, that detailed the history of the DeKalb County Soil Improvement Association.[26] With so many eyes focused on DeKalb, Eckhardt and his University of Illinois sponsors were eager to make the new advisor's transition to DeKalb, and the entire county agent experiment, as successful as possible. In March, thanks to the intercession of his University of Illinois patrons, Eckhardt announced that DeKalb County was moved to the top of the list for a statewide soil survey. Although he was already familiar with DeKalb County soils from his previous research and farmers' institute work, this new survey would be based on some 40,000 individual soil tests and provide the new county advisor with detailed data that, in turn, he could use to prescribe the most effective soil-management techniques. Eckhardt and his supporters were also aware that the prestige of the new soil survey might enhance his reputation with local farmers.[27]

With Eckhardt scheduled to arrive for work on June 1, the Association met in Sycamore on May 11 to formally elect officers and a new executive committee, approve the Association's constitution and bylaws, and generally finalize the governing structure for the new organization. Brown was selected as the president of the Association's board of directors. The close link that the Association made between economics and education were clearly outlined in the new bylaws, which, for example, stated that its objectives were "[t]o promote a more profitable and more permanent system of agriculture . . . [t]o encourage the dissemination of agricultural information . . . [and to] supervise, direct and assist the activities of the DeKalb County Agricultural Demonstrator." Moreover, the Association reaffirmed its position as an organization for all interested DeKalb residents, not just farmers, by opening Association membership to anyone who paid the $5 minimal annual dues. The composition of the board of directors, drawn from the leadership of farmers' institute, the bankers' and newspapermen's associations, and representatives from each county township, also underscored the fact that the Association was not exclusively a farmers' group.[28]

When Eckhardt began work on June 1, 1912, DeKalb County Soil Improvement Association members were excited to see the project finally off the ground. The *DeKalb Daily Chronicle* declared simply that the "big movement which can be said to be paving the way for countless other communities in this country as well as the national organization for soil improvement, can be considered as on the move." Yet, having just avoided the immediate financial pitfalls and won considerable community support for the new advisor in theory, managers of the Soil Improvement Association were cautious. Despite all of their advance planning, the county agent program was an experiment, and nobody wanted to rush blindly ahead and risk the entire enterprise. They were particularly nervous about how the community would respond to Eckhardt in person. There was no way of predicting whether community support for the advisor, which rose steadily during the spring of 1912, could be sustained long enough to ensure the long-term success of their educational experiment. "The work of the demonstrator will, of course, have to be that of a pioneer," cautioned the *Chronicle*'s editors, "as no place in this country has a similar undertaking ever been launched and the DeKalb County men will have to feel their way until the best method of operation can be figured out." There were other questions as well. Although it was apparent that Eckhardt would use the storefront office that was being prepared in DeKalb and would need a car to work the countryside, no one was certain whether he would spend more time in the office or in the field. How much of his time would be spent teaching, writing, speaking, and attending conferences? Would DeKalb farmers even ask Eckhardt's advice? When Eckhardt and the executive committee of the Soil Improvement Association met for the first time, at the Pierce Bank in Sycamore, all agreed that they would need to progress slowly regarding Eckhardt's day-to-day duties until the demands of the job became more clear.[29]

Such apprehensions, however, proved to be unfounded. In fact, the Soil Improvement Association executive committee quickly realized that it faced just the opposite problem, as DeKalb County farmers swamped Eckhardt with requests. On Monday, June 3, his first full day on the job, Eckhardt received six telephone calls asking for his advice. Three days later, he was already visiting local farm fields, answering questions about crops, conducting soil tests, and demonstrating the efficiency of soybeans as a forage crop.[30] He also announced that he was compiling a mailing list of all county farmers for the announcements and bulletins he was planning to write and distribute. Once the executive committee secured a Ford Model T runabout for Eckhardt in mid-June, he could "scurry to every corner of the county on short notice." On June 24, the *DeKalb Daily Chronicle* printed Eckhardt's first bulletin, which advised farmers to write off their failing timothy hay crops and put up oat hay for their horses instead. In late June, Eckhardt was in the southern part of the county, where support for the advisor had always been the weakest, urging farmers to consider new agricultural methods. He also played host to a booster committee that toured the area to drum up support. His press announcement of July 2 invited area farmers to his demonstration plot on the E. H. Olmstead farm, near Genoa, to see for themselves the beneficial effects of applying phosphate to a clover crop. On July 18, local papers ran Eckhardt's next advice column recommending alfalfa as a good crop for improving soil conditions. The steadily increasing demands on the new advisor proved nearly overwhelming—by July, Eckhardt's appointment book was filled through September. County farmers who recognized his car would flag him down for advice, regardless of whether they had an appointment. Eckhardt was also busy with speaking engagements outside DeKalb County to educate other counties about the benefits of a farm advisor program. In fact, he was in such demand that the Soil Improvement Association executive committee agreed with Eckhardt's suggestion that his speaking engagements outside the county should be limited to no more than twelve events a year and that Saturday afternoons be reserved strictly for office calls. The *Prairie Farmer*'s August 1 issue reported on Eckhardt's accomplishments. "The best part of it all," it noted, "was that no one had an axe to grind. It was unselfish, patriotic work, which accounts for much of its success. County demonstration work that is backed by men whose chief thought is to sell something can never attain the fullest measure of success." The county agent program in DeKalb was off to a fast start.[31]

Eckhardt usually spent most of his days during that first summer in DeKalb County walking across fields and dispensing advice. He showed farmers the problems associated with corn root lice and how to apply biphosphate of carbon to prevent skunks from ruining a meadow. He recommended improved farm-management techniques. He rushed to "emergency" calls for crops that were failing. He reminded farmers about the importance of saving only their best corn for the next year's seed stock, and to treat that seed corn with the utmost care. "When the corn is thoroughly dry and before hard freezes, remove the corn to a furnace room where it is dry and will not freeze. The best room in the house," he continued, "is none too good for seed corn. . . . [You] will earn bigger wages for the farm by picking good seed corn than any work that can be done throughout the year."[32] Eckhardt also regularly counseled farmers who had listened to poor advice and thereby endangered their fields. "Remember," he wrote, "the farm is the basis of production, and when you impair its production you destroy the basis from which true wealth is created." He railed against patent stock feeds, quack veterinary medicines, and false advertising. Farmers, for example, who applied burned lime to fields, a commonly advocated folk remedy to improve crop yields, in fact ran the risk that the burned lime would solidify the soils, destroy the humus and other vegetable matter of the soil, and cause serious long-term damage. Eckhardt, like many agronomists at the time, was equally adamant about the careless application of commercial fertilizers that often did more harm than good. Throughout the late summer, he also laid plans for the upcoming autumn and winter. He was invited to speak before the Illinois Bankers' Association, one of the leading proponents of the farm advisor program in the state, in September and needed to prepare his remarks. Other plans included a small greenhouse in his office and regular meetings in local schools, complete with stereographic slide shows, to reinforce the practical farm-management lessons he presented nearly every day as he completed his rounds.[33]

But above all else, Eckhardt was a soils expert, and the Soil Improvement Association entrusted him to restore and maintain the fertility of area farms. Consequently, the most common service he provided was soil testing. Armed with a small soil auger, a box of litmus paper, and vials of hydrochloric acid, Eckhardt found that some DeKalb-area farms had alkaline soils, but the majority of county farms lacked sufficient nitrogen and were too acidic to raise the most successful crops. More ominously, his relentless testing confirmed that decades of farming had taken a great toll on the land and that the most obvious way to improve the soil was to change the common patterns of crop rotation. For most of the county soils, he advocated a cyclical field-rotation program that included commercial crops, legumes, and clover instead of the more common timothy grass, and the careful application of barnyard manure and naturally occurring minerals, such as ground limestone and rock phosphate. He wrote and lectured about the benefits of raising alfalfa. Eckhardt regularly directed farmers to apply limestone to neutralize the field and "sweeten" the soil and then plant clover to increase nitrogen levels. All of this, of course, was closely adjusted to the individual needs of the soil. A typical soil prescription read, "Limestone, 2 tons per acre anytime this fall or winter. Follow in the spring with 14 pounds of clover seed per acre. When you get the soil sweetened and the clover to supply the nitrogen, you won't be satisfied with anything less than 75 or 80 bushels of corn to the acre every year. By that time, you will be ready to put on some rock phosphate, and after that 100 bushels won't be too much to expect."[34]

Farmers had little difficulty securing limestone for their fields. Much of northern Illinois lies over thick layers of limestone, some hundreds of feet thick, which is easily and economically quarried in convenient locations. Limestone for local DeKalb County fields, a byproduct of the construction industry around Chicago, was easy to buy in nearby Elmhurst for only 25 cents a ton. Eckhardt's limestone prescriptions were so frequent that, in early July 1912, the Soil Improvement Association formulated plans to order limestone in bulk and pass the savings on to the local farmers. The Association also arranged for a local fabricator to build a machine capable of spreading crushed limestone and phosphate onto county fields.[35]

Clover seed, however, was another matter. Illinois in 1912 had weak seed quality laws, which meant that the state became a dumping ground for inferior and impure seed stock rejected by neighboring states. Eckhardt knew that such seed, which had low germination rates or was full of noxious weeds, could do DeKalb farmers more harm than good. Moreover, the Association directors realized that crop shortfalls or failures attributable to poor seed might compromise their entire fledgling farm advisor program. Consequently, on July 23, the board of directors of the Soil Improvement Association made their boldest move since hiring their soils expert. At a secret meeting, they directed Eckhardt to buy high-quality clover seed in neighboring states and move it to DeKalb for distribution to area farmers.[36]

The board members were anxious to keep the buying program quiet because they had so much at stake. Clover seed was in high demand across the nation, thanks in part to the Progressive-era farm advisors' new gospel of crop rotation, and board members feared that Eckhardt's buying spree might cause sellers to inflate their prices even more. In addition to authorizing Eckhardt to shop for seed, many members also agreed to endorse promissory notes totaling $20,000 for the purchase of the seed. Small counties, however, hold few real secrets for long, and, in any event, it was difficult to conceal the fact that something important was taking place, especially when the conscientious and easily available Eckhardt cleared his schedule for several weeks in the fall of 1912. The press was even more mysterious. "From the nature of the enterprise," reported the *DeKalb Daily Chronicle,* "it will have to be kept a secret for several months yet but the Chronicle feels warranted in assuring the farmers that it is one of the biggest things ever pulled off hereabouts for their benefit."[37] For months during late summer and early autumn rumors circulated around the county. "The Association, quietly and unobtrusively, has been conducting a campaign in the interests of the farmers of the county which cannot yet be made public and has taken thousands of dollars to finance," the *Chronicle* reassured its readers in September. "As soon as safe the Chronicle will tell of this great work and of other plans which are now in embryo."[38] Still, the Association conspirators kept quiet.

Finally, in late October, the Association announced that Eckhardt had personally supervised the testing of Wisconsin clover seed and bought 1,370 bushels of the high-quality seed—nearly a railroad car full—for about 30 percent less than the prevailing Illinois prices for inferior seed. The bulk purchase saved farmers between five and ten dollars a bushel and proved so popular that farmers placed more than 500 seed orders with the Association. After the original load had been sold, Eckhardt bought another 200 bushels to meet the demand. Still, many seed orders that first fall went unfilled.[39]

The success of the cooperative limestone purchase and the seed sale underscored Eckhardt's ability to convince local farmers to adopt modern agricultural methods. It also showed the importance of patient personal interaction between the farmers and their advisor as part of sustainable agriculture in the county. More important, it proved that cooperative purchasing of farm supplies provided farmers with high-quality products at below-market rates, a lesson that the Association would remember in the future. In fact, one strong indication that the "Soil Improvers" were committed to cooperative seed purchasing in the future, and perhaps even hoped to expand their efforts, was that the Association invested considerable money in machines to clean clover seed and a moisture tester for corn seed.[40]

At the end of 1912, directors of the Soil Improvement Association met with Eckhardt to assess the first year of the county farm advisor program. They were justifiably proud of their accomplishments. In less than a year, the Association had secured widespread and diverse community support, raised money, convinced skeptics, and succeeded in hiring one of the leading soils experts in the state as their county agricultural advisor. Its experiment in bulk purchases of limestone demonstrated the economic advantages of collective buying. The clover-seed program provided area farmers with a rich stock of seed that would improve area soils. Even by the end of only the first growing season, farmers could see tangible results from Eckhardt's ongoing advice—the demonstration plots at the DeKalb County farm, for example, were growing impressive stands of corn and a new, exotic crop called "soybeans." Perhaps most important, Eckhardt had won the confidence of county farmers eager to boost their yields and farm profits and been warmly received by the business community, political leaders, journalists, the county's leading citizens. The *Breeder's Gazette* summarized how many DeKalb citizens felt about their new "county agriculturalist":

> There are some men who can command confidence, respect, and obedience. It is a tremendous undertaking, this idea of a demonstrator going into a county, perhaps strange to him, to study its soils, crops, animals, men, women, children, to have visions and faith to plan out better ways, not merely temporarily better ways, but permanently better ones. It needs a man with a lot of agricultural science, a thorough grounding in agricultural practice, an economist, a preacher, an exhorter, a man of patience, faith, hope, and kindling enthusiasm. When you see such a man as that going down the road, whether in an automobile or afoot, stop him, pull him in, let the children and the wife meet him.[41]

The DeKalb experiment in community agricultural education was praised by leading agricultural journals. "A few far-sighted men behind the scheme," lavished Clifford V. Gregory, editor of *Prairie Farmer*, "had a dream of doubled crop production—of prosperity resulting from these bountiful yields that should make DeKalb County the best place in the world to live. . . . The best part of it was that no one had an axe to grind. It was unselfish, patriotic work, which accounts for much of its success. . . . It is epochal, this work that DeKalb County is doing. With unselfish courage and high ideals the people of DeKalb are blazing a new trail—a trail that will lead to undreamed of prosperity, a trail that will be eagerly followed by other counties as their eyes are opened to the way of agricultural salvation."[42] It ran photos of the S. A. Jones farm in Shabbona, which had produced poorly prior to Eckhardt's prescription for $10 worth of potassium chloride but yielded nearly 70 bushels per acre after treatment. These were results that framers could readily understand. *Prairie Farmer* also ran Eckhardt's articles. *Breeder's Gazette* noted,

> There is something new in the land. A man, clad with no authority to compel, but armed with a knowledge of good farm practices, goes about his county, counselling this man and that to reform his ways, to forsake his slipshod or erroneous farm practices, feel change of heart and help in the great movement for farm uplift. This man is the county demonstrator. . . . In rich counties, such as we see in Illinois, he is provided with automo-

bile and stenographer, he has his office in the court house whence he sallies out on missions of good import. . . . He is a man supposedly full of good practical ideas, and it is his mission to so modify the farming of his county that he will have earned his salary and a good deal more.[43]

The successful DeKalb County experience also served as a model for other counties to organize and hire their own agents. In October 1912, the University of Illinois appointed a committee to develop "farm expert" guidelines. The resulting procedures for establishing a county organization were clearly influenced by the lessons learned in DeKalb. Once 300 farmers in an Illinois county requested assistance, for example, they could form an organization to cooperate with the College of Agriculture and, after passage of Smith-Lever in 1914, with the U.S. Department of Agriculture, in supporting and managing a county farm advisor. Each county organization was required to raise money sufficient to finance at least half of the costs for an advisor. Advisor candidates would be approved by the college only after showing convincing evidence of professional training, five years of agronomy practice, and practical farming experience. The DeKalb prototype of a local organization working with the University of Illinois to select a farm advisor, and providing that advisor with salary, office, staff, and equipment, proved to be effective—by 1921, all but seven Illinois counties had a local farm advisor organization modeled largely on the DeKalb model. Even the Soil Improvement Association's financial struggles provided valuable lessons. "The biggest factor in making this county demonstration work a success," wrote Eckhardt, "is that a large number of local people be interested. . . . [c]ontributions may be from $1 up; probably $25 being about the maximum amount you will be able to get from any one farmer. It is best to have these amounts contributed on a yearly basis— that is, if a man subscribes $10, let that be $10 a year for three years."[44]

Beyond the tangible benefits for farmers, DeKalb County's experiment in agricultural education and services available to all DeKalb County residents regardless of whether they belonged to the Soil Improvement Association reflected Progressive ideals that stressed support for agriculture, community reform through education, reliance upon a trained expert to diagnose and solve seemingly intractable problems, and the inherent nobility of American farmers. Yet, as much as it mirrored many national agricultural reform trends of the era, the DeKalb County experience is equally noteworthy for its departure from the period's prevailing currents of modernization. A great gulf separated the two leading philosophies about agricultural advisor work. Many Progressives, especially in the federal government and land grant colleges, favored a "top-down" approach by which federal agents worked directly with farmers. This ideal is most clearly reflected in the national consensus that resulted in the Smith-Lever Act in 1914. The Soil Improvement Association, on the other hand, was a local affair that relied for its success upon strong-willed, self-confident, internal leadership from social and economic elites within the community supported by broad popular participation from area farmers. Although it drew energy from the same reform forces that resulted in the national Smith-Lever Act, DeKalb County leaders acted on their own to establish a county agent program months before Congress passed its own federal county agent law. DeKalb County leaders acted with little direct financial support from either state or federal governments to provide local farmers with a full-time agricultural expert who would ensure the long-term prosperity of the county.

DeKalb County farmers historically enjoyed considerable financial success, so their experiment in adult agricultural education was directed toward teaching modern agricultural methods that would sustain their prosperity. Consequently, many county farmers were eager to adopt the new techniques. Most important, the Soil Improvement Association was ultimately built upon the cooperation between farmers, bankers, newspaper editors, merchants, clergy, and educators. What bound these segments of the local community so tightly together was their common interest in preserving the county's most treasured asset, its native soil. The Soil Improvement Association was, in the end, the child of that collective community spirit.

3 WAR AND RECESSION

EARLY TRIALS FOR THE

SOIL IMPROVEMENT ASSOCIATION

• America was in its "Golden Age" of agriculture when the leading citizens of DeKalb County undertook their experiment in adult agricultural education. The nation was adjusting to the rapid commercialization and mechanization of agriculture that occurred during the closing decades of the nineteenth century and struck a tentative balance between its agricultural and industrial economies. Commodity prices were relatively stable because, although the nation's farms continued to produce at high levels, they were not pouring out food and fiber at rates that far exceeded domestic and international demand. Europeans and other international buyers provided markets for some of the excess output. The farm economy seemed able to absorb minor fluctuations in commodity prices, and, in general, farms were profitable, especially in the upper Midwest. The country was laced together by an extensive railroad network, and, although it was not always efficient and reliable, it nevertheless provided transportation links to even the most remote farming regions. Cars and government-supported road construction were improving living standards down on the farm. Trucks, which were beginning to appear on farms and would become a standard feature of American agriculture after World War I, were poised to revolutionize agricultural marketing techniques and open new economic opportunities for rural folk. Expanding industrial production demanded ever-increasing amounts of labor, and that labor force was not highly unionized. Consequently, shifting from one occupation to another was fairly easy, making industry an attractive outlet for the natural surplus of farm children who had diminishing hope of acquiring their own land in adulthood. In most areas of the country, rural education reached a level where many native-born, white rural children learned academic skills that prepared them for industrial work and city living and, sometimes, even made them competitive for urban professional jobs. Social institutions, such as churches and clubs, played a vital role in maintaining a strong sense of community among rural Americans. Work hours on the farm were long, but laborsaving devices of all kinds eased the burdens for both farmers and their families, increased production, reduced livestock and labor costs, and enriched the daily lives of farmers. The demand for productive land was strong, which suggests that farming continued to be an attractive alternative, economically and emotionally, to other economic opportunities.

The county that welcomed William Eckhardt to his new job as farm advisor embodied much of that economic success.[1] Thrift, hard work, Christian morality, Progressive-era confidence, and good fortune had indeed led to prosperity for most DeKalb County farm owners—a good life, for which earlier generations had struggled so hard, had become a realistic expectation. DeKalb County had just under 2,500 farms, valued at roughly $58 million. Farms in DeKalb County were worth, on average, about $23,000, an amount between $4,500 and $9,500 more than average farm values in eight surrounding counties.[2] DeKalb farms sold for between $150 and $200 per acre, with the most desirable land appreciating at nearly 15 percent per year.[3] Crop and livestock prices always fluctuate, but in the spring of 1912 the trend was clearly toward prices that were higher than they had been the year before, and prospects were good for continued favorable returns. Farm labor was scarce and wages stood at around $2.50 per day.[4] The county's twenty-one banks were secure, with combined assets of more than $4.8 million. Credit for farm improvement was readily available.[5] Farming itself was changing dramatically, and Eckhardt was brimming with optimism as he began his new position as county advisor for the DeKalb County Soil Improvement Association. Area farmers who followed Eckhardt's advice were confident that they were employing the latest agricultural techniques to ensure their continued success.

The prosperity and confidence, however, did not translate into financial security for the Soil Improvement Association during its formative years. Paying for the farm advisor enterprise was a greater problem than anyone in the Association leadership admitted publicly. The local bankers honored their pledges to contribute $2,000 for the first three years, and the DeKalb County Board of Supervisors still had Eckhardt on its payroll for $2,000 as a county employee, leaving $6,000 of the remaining $10,000 for the year to be raised by new Association subscriptions. The county community was delighted with Eckhardt's work, but the Association was not collecting sufficient membership fees to offset its costs. About 700 farmers subscribed during the first year to support the Association, but the townships in the southern part of the county were behind in suggested membership quotas, and the new season of advisor work loomed with little prospect for new sources of revenue.[6]

DeKalb County was not alone in having trouble funding its farm advisor. As the farm advisor movement spread throughout the state, other counties had as much or more difficulty in raising money to support a farm advisor. Consequently, during the winter of 1912–1913, the Illinois state legislature began debating a proposed law to permit county boards of supervisors to appropriate up to $5,000 a year toward the maintenance of a county agricultural agent. The possibility of greater public assistance caused Soil Improvement Association leaders to consider the future of their new organization. They were torn between two inconsistent desires—their immediate need for increased and predictable public revenue that the law provided and their steadfast determination to remain free from outside influences. The Soil Improvement Association's ideal of employing its own advisor, and seeking technical or management assistance from specialists or governmental institutions when the need arose, was rooted in their traditional democratic and agrarian ideals of local control and self-help. Association leaders were convinced that farm advisory work was most likely to succeed in the long run if local people defined the goals for their organization and took the primary responsibility for planning and carrying out their own programs.[7]

On the other hand, although many believed that the advisor's work was a public service that should be supported by taxes, Association leadership did not wish to be obligated to a political organization. The Association, after all, had been conceived and organized by local supporters and was free from cooperative agreements, with the exception of the arrangement it already had with the DeKalb County board to pay $2,000 of Eckhardt's salary. Nevertheless, their need for revenue and determination to see their new enterprise succeed led most Association members to support the proposed legislation cautiously. When the law passed, in June 1913, most Association members were relieved that their ad hoc financial relationship with the county was now sanctioned by state law and that they could look to the county government to provide as much as $5,000 each year. More important, although they remained cautious about entangling the Association too deeply in local or state political affairs, they were pleased that the state legislature recognized the need for public funds to support farm advisors while enabling the Association to maintain control of its programs.[8]

At the same time the Illinois law was working its way through the state legislature, Congress was debating the Smith-Lever Act, a law that, in part, provided $1,200 in direct federal assistance a year to county organizations that employed an approved county agent. Smith-Lever became law in 1914, and in early October the Soil Improvement Association's board of directors applied for the new federal funds. Agreeing to the terms of the Smith-Lever Act, however, placed the Association in administrative limbo. The federal law created a framework for the establishment and oversight of county advisors through state land-grant colleges and the newly formed Cooperative Extension Service. In Illinois, for example, the administrative agency was the University of Illinois, which appointed Dean Eugene Davenport as the first director of the Cooperative Extension Service, with W. F. Handschin as the vice director and state leader. DeKalb County's program, however, was created before the passage of the act and owed its existence to the tireless efforts of the county's own leaders. They were jealous of their independence, proud of their success, and suspicious of big government involvement in their local affairs. Consequently, the Association's leadership was unwilling to give up its autonomy and turn control of the advisor program over to the Cooperative Extension Service. In the end, the Association and the Extension Service reached a compromise. The county would accept Smith-Lever benefits and submit its personnel and programs to general review as required under the law; however, it maintained its independence and operated in conjunction with, but not within, the federal extension service organizational framework. It was an informal arrangement that clearly benefited the Soil Improvement Association more than the Extension Service—the private organization could collect public funds and still maintain its independent status, freedom of action, and local control.

The new relationship brought in needed revenue, but Association leaders also understood that the political independence of their organization rested squarely on their ability to generate a high level of local interest in the programs and raise money within the county. In addition, although the Association had spent considerable sums to purchase equipment and rent workspace, it still needed a steady inflow of money for routine expenses, and even greater amounts for Eckhardt and the Association to expand their services. To meet those needs, Association spokesmen worked constantly during the early years of the Association to drum up interest, and solicit money, from area farmers. Henry Parke, for example, worked to convince farmers to support the Association by reminding them about the practical reasons to join. He noted that state livestock officials had released DeKalb-area farmers from hoof-and-mouth disease quarantine in 1915 some sixty days earlier than adjoining counties because of the efforts by Eckhardt and the Association to certify the county's herds as disease-free. He pointed out that education had led to proper crop rotation to reduce insects and disease, which in turn, boosted yields and profits. Consumption of soil additives was up and more farmers were planting clover to help maintain soil fertility, and the costs for those improvements was down, thanks to the Association's collective buying programs. In a bid to increase long-term membership to at least 1,000, the Association also lowered dues from $5.00 to $2.50 annually and asked for a three-year membership commitment.[9]

Although the "Soilers" struggled financially and wrestled with philosophical and practical management details during the early years, they were careful not to let those concerns interfere with Eckhardt's day-to-day operations, and, from the outset, their experiment in adult agricultural education was popular among DeKalb County farmers, if not always well supported. Demand for expert advice was always strong. During the winter of 1912–1913, after the 1912 harvest was complete, Eckhardt spent time fielding telephone inquiries, making speeches, helping with the farmers' institutes and township farm clubs, and working in the laboratory, which included a new suite of equipment for testing seed. Eckhardt also wrote and published a farm advice column that appeared periodically in local newspapers. More than anything else, however, Eckhardt was swamped with requests to prescribe soil improvements. The success of his recommendations the first summer, evident to anyone who inspected the treated fields and the county demonstration farm, generated excitement among farmers about the "clover campaign" that promised to restore the soil and fill farmers' pocketbooks. More county farmers than ever wanted to recondition their soil. By the middle of March 1913, Eckhardt's time was already booked for the next two months with 139 appointments.[10]

Eckhardt campaigned relentlessly for farmers to rotate their fields, plant more clover, and consider other crops, including alfalfa and soybeans, as ways to maintain soil fertility. He also counseled farmers to use additives, such as limestone and rock phosphate. Whereas limestone was readily available, rock phosphate needed to be purchased from dealers and transported by rail to area farms. To meet the increasing demand for phosphate, the Soil Improvement Association instituted a program in 1913, modeled on its successful limestone and clover seed programs the preceding year, to buy and ship rock phosphate in bulk and pass on the savings to area farmers. The Association announced in August, 1913, that it would again buy clover seed for the county. Eckhardt also regularly cautioned his constituents that the commercial fertilizers available at the time often caused more harm than good. "We hope no DeKalb County farmer or landowner will ever buy a dollar's worth of fertilizer," Eckhardt warned, "without knowing absolutely what he is doing. . . . Without a proper knowledge of what to do, a farmer may ruin his land, as past experience only too clearly shows." To protect farmers, Eckhardt instituted a new service to analyze fertilizers to save farmers from wasting money, or even damaging their fields, with the inferior-quality fertilizers that were commonly sold to unwary farmers.[11]

The soil expert was equally busy off the county fields with office calls, experiments, and writing. By the middle of September 1913, Eckhardt's calendar was full through the end of the year and he was already booking appointments for 1914. He was also becoming a leading spokesman on the farm advisor movement regionally and attended many speaking engagements to promote soil improvement. During the summer of 1913, for example, Eckhardt hosted visiting delegations from other counties who sought his advice on how to establish their own farm advisor programs. He spoke, along with Congressman A. F. Lever of South Carolina and the president of the Rock Island Railroad, H. U. Mudge, at the Union League Club in Chicago to outline the success of DeKalb's program and encourage other counties to establish their own farm advisor organizations.[12]

During the winter of 1913–1914, Eckhardt and the Association concluded that, while two summers of field testing and hands-on demonstration had worked productivity miracles on many DeKalb fields and was clearly vital to the continued success of area farms, they needed to do still more. Eckhardt suggested that the Association provide members with a regular periodical that would help keep farmers informed and engaged in the farm advisor program. In February 1914, despite the fact that the Association was still struggling financially, it approved the publication of a monthly bulletin, the *DeKalb County Farmer*, with Eckhardt serving as the editor, business manager, and "Consulting Agriculturalist." The first edition appeared in farmers' mailboxes the next month. The official aim of the publication was to inform Association members, through articles and photographs, about agricultural news and the latest practical advice that farmers could follow. "Our greatest efforts and service," wrote Eckhardt, "has been [sic] to constantly present those facts worked out by science that will help bridge the gap between farm practice and a permanent and profitable agriculture. We publish a paper, the *DeKalb County Farmer*, in which we constantly keep the fundamental principles before the farmers of this County." Unofficially, they also hoped that such a publication would foster a stronger sense of community among members of the Soil Improvement Association, attract new members and funding, and, in general, help maintain peak interest in the farm advisor program.[13]

In addition to recommending sound soil practices, offering testing services, teaching area farmers the latest agricultural techniques, giving public lectures, and writing articles for the *Farmer* and other agricultural journals, Eckhardt was researching new varieties of corn seed he hoped would provide still higher yields. There was a limit, after all, to what sound soil practices could achieve, and DeKalb County fields would never reach their full potential unless farmers planted improved varieties of crops, especially corn. At the time, however, commercial seed was so expensive, difficult to acquire, and inconsistent in quality that few farmers bought enough commercial seed to plant all of their fields. Most, instead, simply handpicked the very best ears from the open-pollinated corn in their fields as their seed stock for the following season. Eckhardt, in fact, regularly reminded farmers about how to properly choose, prepare, store, and test such open-pollinated corn seed. But Eckhardt was also one of a growing number of agriculture specialists across the nation who believed that improved yields from laboriously and meticulously bred seed stock would be enough to offset the seed purchase costs and return handsome farm profits. One variety that attracted him was Western Plowman corn seed, perfected through breeding and careful harvesting in nearby Will County. Generations of painstaking selection had resulted in a corn plant that was perfectly suited to DeKalb's soil, climate, and harvesting techniques. The results were encouraging. Experimental plots of Western Plowman in DeKalb County produced between 3.7 and 19 more bushels per acre than the common field corn popular with farmers. It showed such potential that, in the autumn of 1914, Eckhardt and the Association stockpiled 3,200 bushels of Western Plowman for county farmers, enough to plant about 15,000 acres. This corn seed promised between 4 and 20 percent yield increases—if all corn fields in DeKalb had been planted with this variety to increase yields by only 4 percent, the county might earn around $250,000 more total farm income, or an average of about $100 per farm.[14]

DeKalb's farm advisor was not only deeply committed to working for his constituents, but was also a leading figure in the evolution of farm advisor programs statewide. Progressive-era interest in practical agricultural education, combined with the success of DeKalb's experiment, led farmers and rural businessmen in other Illinois counties to consider forming agricultural associations and hiring their own farm advisors. The new Smith-Lever Act generated even greater momentum—by mid-1914, fifteen Illinois counties, mostly in the northern portion of the state, had their own farm advisor programs, and many more were in various stages of organization.[15] Demand for qualified men was growing steadily, and county associations understandably preferred to hire Illinois men who were most likely to be familiar with the state soils. With salaries as high as $4,000 annually, there was no shortage of volunteers for advisor jobs. The requirements the University of Illinois set for agricultural advisors, in its capacity as the oversight body for the county advisors under the Smith-Lever Act, were, however, strenuous, and the university was highly selective. Probably the most difficult hurdle for applicants was the one that required farm advisors to have five years of practical experience in agricultural science and education after graduation from college before they could even be considered for an official advisor position. Such high standards ensured not only that county advisors were educated and well trained, but also that they would be professionals who could earn and keep the respect of their constituents.

With demand for agriculturalists growing steadily throughout Illinois and much of the nation, advisors and leaders of county agricultural associations became concerned that the strict qualification requirements for candidates might be compromised in order to produce sufficient numbers of advisors to fill the burgeoning demand, thereby lowering the quality of the agricultural services and possibly endangering the reputation of all farm advisor programs. Rumors even spread that some Illinois counties were considering doing without the approval of the university and U.S. Department of Agriculture (USDA) and hiring men who were not even university graduates and were poorly trained in practical agriculture.

The shortage of qualified candidates was complicated by the national debate about how best to provide agricultural education. Two opposing ideals emerged. On one hand, many commentators favored a decentralized model, such as the one pioneered by DeKalb, in which farmers and other community leaders organized and supported their own county advisors, perhaps with indirect governmental assistance, sought advice and supervision from state land-grant universities when the need arose, and had nearly total control over the affairs of the Association. On the other side of the debate there was strong support for a centralized advisor program, run largely by the USDA, that would put federal agricultural agents in the fields to assist farmers. Eckhardt and the Soil Improvement Association, and advisors and support organizations in other Illinois counties, were philosophically opposed to this second alternative. County advisors, associations, and scholars at several land-grant colleges, including the University of Illinois, feared that too much intrusion by the federal government into their county advisor programs would undercut their operational freedom and lead to increased bureaucratization, red tape, and regimentation that would hinder rather than assist the development and operation of independent, locally controlled county advisor organizations.

With these concerns in mind, several Illinois county agents, including Eckhardt, began to discuss the need to create an organization of farm advisors who would share information and experiences, maintain and defend high educational and ethical standards, defend their autonomy, and generally ensure the professional nature of their positions. Such a professional organization was not unique at the time. It reflected, instead, a broader Progressive-era emphasis upon organization of professionals in many fields, such as medicine, engineering, and law, for the purpose of articulating a collective opinion on important issues. Moreover, such professional organizations used their prestige as the embodiment of expert knowledge in their fields to influence public policy.[16] It seemed natural to professional agriculturalists like Eckhardt that organizing into an association was one way to ensure the quality of advisor candidates, influence farm policy, and preserve their vision of the farm advisor program. Consequently, in late 1913, although they were still a small fraternity, the farm advisors created the Illinois Association of County Agriculturalists (later changed to "Advisors"), the first association of county extension workers in the nation. The new organization elected Kane County farm advisor Jerome E. Readhimer as president and Eckhardt as vice president.[17] At their winter meeting in January 1914, Dean Davenport praised their efforts. "I am glad," said Davenport, "that the county advisors of Illinois have formed an association for their advice and benefit. It should operate in many ways to the advantage of your work in the field." By the middle of that year, there were thirteen advisors meeting twice annually; by early 1915, sixteen county advisors had joined the new organization.[18]

The success of the Illinois Association of County Agriculturalists convinced members that they should work to expand the organization to a regional, and perhaps national, level. At the spring 1914 meeting of the new organization, Eckhardt proposed that county agents from other midwestern counties meet during the Chicago International Livestock Exposition the following December to discuss the budding agricultural advisor movement and consider forming a national association of advisors similar to the one recently created in Illinois. Twenty-five advisors from several states gathered at a restaurant on South Wabash Avenue in Chicago for that first meeting of what eventually evolved into the National Association of County Agricultural Agents. Although Eckhardt never served as an officer in the national organization, contemporary observers of the events leading to the formation of both the Illinois and the national associations acknowledged his great organization skills and credited his leadership as one of the most important driving forces during the associations' formative years.[19]

The Illinois Association of County Advisors quickly moved beyond simply sharing practical information among members. At their January 1915 meeting, the membership resolved to support agricultural policy issues beneficial to all state farmers. They advocated, for example, that hog cholera laws be more vigorously enforced, that the state be quarantined against cattle with tuberculosis or other diseases, that all dairy products be pasteurized, that foot-and-mouth indemnities be paid to farmers from public funds, and that the state establish a veterinary college at the University of Illinois. Beyond these practical issues, many in the Illinois Association of County Advisors, including Eckhardt, believed it was necessary for the county associations themselves, such as the DeKalb County Soil Improvement Association, to band together, just as the advisors had done. By the summer of 1915, they became convinced that, as state and federal governments became more deeply involved in agricultural policy making, the associations needed to speak with a united voice on important agricultural issues. Moreover, collective action to purchase quality seed stocks or soil additives could increase purchasing volume and decrease costs to farmers.

The following winter, at their January 1916 meeting, the advisors adopted a resolution to formally consider such a statewide association, to be called the Illinois Agricultural Association. In keeping with the spirit of self-governance that had worked so well in DeKalb County, Eckhardt told his colleagues that the new organization should be composed of farmers and those whose fortunes were tied closely to agriculture, with county advisors offering advice when necessary. The editors of farm journals were jubilant about the possibilities for the new organization of county farm associations. "Who can foretell," asked the *Orange Judd Farmer*, "the power of such an organization of leading farmers of the state with hearts set on rural betterment, and with selfish interests far in the background?" "It is planned," wrote the *Prairie Farmer*, "to make the Illinois Agricultural Association a strong, statewide organization that will have the strength and the purpose to work for the interests of Illinois farmers in every possible way. There has been a great need of such an association in Illinois and farmers' interests have suffered for it. . . . One of the most powerful effects of this association will be its wholesome influence on Illinois politics. . . . Let's all work together to make the Illinois Agricultural Association the greatest constructive force in the state." In March 1916, delegates to the Illinois Agricultural Association organizational meeting approved the bylaws and selected permanent officers, and thirteen Illinois county associations joined. From that modest beginning, the Illinois Agricultural Association became, and remains today, one of the most influential state farm bureaus in America and an influential voice in the development of national agricultural policy.[20]

Eckhardt's star was rising fast, and by the middle of 1916 he had emerged as a leading voice among the advisors in the Midwest. His main focus of responsibility, however, remained the farmers in DeKalb. As required under the terms of the Smith-Lever Act, beginning in 1915 Eckhardt submitted statistical and narrative reports to the federal Cooperative Extension Service that answered specific USDA inquiries. In general, the report reviewed the yearly accomplishments of the county and advisor. These accounts also reveal much about Eckhardt and his philosophy about farm life. He was especially pleased with the way in which county farmers embraced new crops and soil additives that helped sustain soil fertility and boosted production. Of all the Association's activities, however, farmers most appreciated its willingness to obtain high-quality seed. Eckhardt noted, for example, that alfalfa seed consumption increased from a few bushels to more than 1,000 bushels between 1914 and 1915, with nearly 15,000 acres in the county planted with alfalfa. Eckhardt also highlighted the Association's program to purchase weed-free clover seed, Western Plowman corn, alfalfa seed, and oat seed—in total, during 1915 the Association and a handful of its wealthiest members provided $33,000 worth of high-quality seed almost at cost to local farmers; a year later, the Association stockpiled $35,000 worth of seed. Moreover, small handling fees from the sale of the seed and soil supplements helped in stabilizing the Association's finances.

DeKalb County was a crop-farming county, but like most corn belt counties, it also had several large livestock operations as well as many individual farms that raised livestock along with grain. Eckhardt, although trained as a soils expert, offered his services and advice to the stockmen as best he could. Much of it was fairly simple. "First we encourage every farmer," Eckhardt wrote, "to grow one-fourth or nearly so of his cultivated land in clover or a relative crop. The plan being to use this crop on the farm to produce milk, beef, or pork."[21] He was far more effective in his role as a spokesman for livestock interests. "Every effort possible has been used to encourage better livestock," reported Eckhardt. "We expect or hope to form an Association of all pure bred livestock raisers in order that their stock be better known within the County as well as furnish a market for the surplus."[22] Eckhardt relied heavily on the advice of local veterinarians who recommended that county stockmen who had been hit hard by hog cholera and hoof-and-mouth disease work with the Association to implement quarantine, isolation, and sanitation programs. Such timely measures prevented the further spread of those diseases and returned DeKalb herds to production faster than other counties.[23] Eckhardt also worked to educate stockmen about worthless patent medicines that were widely available to gullible farmers. "Do not blame the farmer, wrote Eckhardt, "for taking a chance [with patent medicine]. No stock has paid off as many mortgages nor caused as much worry and uncertainty as hogs. The scoundrel that will induce a farmer to use worthless remedy and lose his herd of hogs, in other words, a year of hard labor, deserves a term at Joliet under state supervision."[24] Eckhardt and the Association, in fact, were instrumental in exposing a patent hog medicine fraud that led to an arrest warrant for one notorious veterinary quack—"parasite" in Eckhardt's words—operating under the name "Dr. D. W. Nolan."[25]

The report also revealed much about Eckhardt's underlying philosophy toward farm advisor work. He was outspoken, for example, about working with children. In response to the Extension Service inquiry about his work with local boys' and girls' clubs, he reported, "In my judgment if a County Agent has no better services to contribute to his County than fool his time away with Boys' and Girls' Club work, the County had better do without his services. Whenever people reach such a low degree of intelligence that they must be taught by their children," he continued, "it will probably be more economical and desirable to have them replaced by other and better stock." He was no more receptive to county home advisor work. Neighboring Kankakee County had recently hired the first county home advisor, Eva Benefiel, in 1915. When asked about hiring a farm home management advisor for DeKalb County, Eckhardt was adamant. "I would say NO! Farm women's labor is being lessened by machinery as rapidly as people can do. The farm woman rears healthier children, as a rule, in a country home and served an apprenticeship with a practical mother fitting her wonderfully for her work when supplemented by proper education. I doubt whether such a project would receive much consideration outside a few city women who feel they are endowed with gifts to save their country sisters."[26]

Eckhardt also appeared to have little sympathy for the county's poorer farmers. Like most rural Americans of the period, he embraced long-standing American agrarian ideals about the fundamental nobility of farm work, the inherent values of rugged individualism, and the inevitability of success brought about by hard work. Consequently, he was far more conscious about highlighting the opportunities for self-help for all the county's residents than to acknowledge the structural conditions that led to rural poverty, whose very existence ran contrary to his underlying assumptions. "The poor tenant farmer who does not have his few dollars for a membership fee," wrote Eckhardt, "is given the same careful consideration and treatment as the man who may be worth a hundred or two hundred thousand dollars. . . . I believe one of the greatest services we have rendered our county as an organization," he reflected, "is by leaving the social side of country people alone. I do not believe in the 'uplifting of the downtrodden farmer.' We do believe that the farmer is abundantly capable of caring for himself and as a class is perhaps better off than any other class of people. Farmers, as a whole, resent exceedingly those forces which are at work with missionary intent trying to uplift them and those men who will undertake this side of country life, will meet with opposition that is justly presented." In another part of his report, he wrote simply, "The people have a pride in their work and are capable of enjoying themselves without outside help."[27]

Likewise, Eckhardt had little patience with the whole reporting process mandated by the USDA under the terms of the Smith-Lever Act and was highly critical of the "great Department of Agriculture in Washington." He seemed to delight in presenting statistical data about programs that he favored, but responded, "We have no time for gathering statistics" to the question about how many farmers practiced various farming practices. Where asked to provide a detailed proportional accounting of how much time he spent in the field, office, conducting demonstration work, attending meetings, and carrying out other requirements of his job, Eckhardt scrawled this sarcastic note: "We will gladly devote our time to do such work when we are so instructed by our State leaders that our time is of no greater value, and such a program is approved by our local Executive Committee and Directors."[28] He was equally clear about the efficacy of the whole reporting process. "Of the 148 questions asked," he reported, "less that half a dozen contribute to the central fact, that is, larger returns for human labor and a permanent and profitable agriculture . . . Such questions and the desire for such data is an insult to any intelligent man fit to hold the position of County Agent." The tone of these responses further illustrates the fundamental gulf that existed between the expectations of the essentially private and independent Soil Improvement Association and the expectations of the federal extension program.[29]

The most persistent message throughout the reports, however, was the success of the Soil Improvement Association's program of purchasing quality seed and soil additives for resale to county farmers. The service was so wildly popular that it seemed at times as if that enterprise would eclipse the other services provided by the Association. By 1915, the Association had expanded its seed enterprise, now called DeKalb County Soil Improvement Association Seed Fund, to include not only clover but also timothy, oats, alfalfa, and seed corn. The business, however, was always a risky financial burden, and the stakes were getting higher. Each year, beginning in 1912, a handful of wealthy and respected individuals guaranteed the notes that the Association used to buy the seed, limestone, and phosphate. The first notes were for a relatively modest $16,000, but the amounts quickly escalated, and by 1916, it looked as if the Soil Improvement Association would need $75,000 in personal guarantees to support its seed business. In addition to purchasing quality seed, Eckhardt was also working with growers to breed and select improved strains of Western Plowman corn seed. It was becoming quickly evident that the Association's seed and additive programs were outpacing Eckhardt's ability to manage the service effectively. Moreover, there was always the underlying fear that a major financial reversal in the seed program might doom the entire farm advisor program. Finally, the USDA Extension Service began to question the propriety of having a farm advisor, who was regulated by a federal government agency, involved in a private business venture like the seed and additive program, over which the government had no control.

In April 1917, Eckhardt addressed a letter to the Soil Improvement Association officers and directors suggesting that they create a corporation to take over the seed-buying program and put it on a sound and permanent financial footing. "It does not seem fair or should it be expected," Eckhardt wrote, "that 30 to 40 men sign a guarantee running into the price of one to two good farms and especially when the benefit to be derived from the service goes alike to anyone of over 2,400 farms that wish to avail themselves of same. The 2,400 farms means an agricultural wealth of stock, equipment, buildings, and land of perhaps $70,000,000. Surely, among this group there are 200 or 250 men who will feel it a privilege and duty to take at least one share of stock at $100 each in an Association to provide a fund of $20,000 to $25,000 and relieve the small group of shouldering the big load." The Soil Improvement Association's directors agreed, and on June 2, 1917, they voted to "form a separate organization to care for the seed business and other business of like nature." On June 21, the Illinois secretary of state incorporated the DeKalb County Agricultural Association, with an initial capitalization of $40,000 raised by the sale of stock. Shareholders paid $100 per share, with a five-share limit, and expected 6 percent dividends. From then on, the DeKalb County Soil Improvement Association was officially out of the seed business. The ties between the Soil Improvement Association and its closely affiliated Agricultural Association, however, remained strong. Indeed, although incorporation changed the legal status of the business, there was little practical change in the way it conducted its affairs, and many of the most important Soil Improvers were deeply involved in the affairs of the new business.[30]

The new corporation freed Eckhardt to devote more of his attention to educating DeKalb farmers about modern agriculture. In fact, the demand for Eckhardt's services and the hopeful prospects of the new seed company convinced the Soil Improvement Association directors that both organizations needed a larger home. The storefront offices in DeKalb were simply too small. The following summer, on July 1, 1918, the Soil Improvement Association bought the old North School in DeKalb for $3,000. In the new facility, the Soil Improvement Association had the room to expand, improve its operations, and work to help farmers sustain the surging prosperity of the era. Four months later, in October 1918, the Soil Improvement Association sold part of the North Fifth Street building and the remaining business assets of the now defunct DeKalb County Soil Improvement Association Seed Fund to the new DeKalb County Agricultural Association for 109 shares of DeKalb Agricultural Association stock. The Soil Improvement Association sold twenty-nine shares of the stock and retained eighty, with a recorded book value of $8,000, although the real value of the shares was probably considerably less. It was clearly a mutually beneficial transaction. The new seed corporation needed offices and storage space but had little cash to spare for capital investment, and the Soil Improvement Association, while also always looking for ways to increase its revenue, had noncash assets to spare for the corporation that was so closely related to the Soil Improvement Association's fundamental mission. In the short run, the deal ensured the continued availability of high-quality seed to area farmers. More important, in the long run it laid the groundwork for the financial success of the Agricultural Association's future hybrid seed business that, in turn, would provide the revenue for the DeKalb County Farm Bureau's extensive educational and service programming during the second half of the twentieth century. It was this initial stock investment that eventually propelled the DeKalb County Farm Bureau to the forefront of the farm bureau movement nationwide.[31]

Many of the local and state activities that made DeKalb County's experiment in adult agricultural education and service a model private agricultural association took place against a backdrop of war. In August 1914, just two and a half years after the founding of the Soil Improvement Association, the nations of Europe plunged themselves into World War I. The outbreak of the war caught most of the belligerent nations woefully unprepared to sustain their massive armies for long periods of combat. The leading European military strategists simply assumed that no nation could prosecute a continental campaign and absorb the levels of destruction that would result from months, much less years, of sustained military operations, using modern weapons, across thousands of square miles of territory and ocean. Consequently, they had predicted and planned for years that any general European war would be violent, short, and decisive—the German General Staff, for example, believed its armies could defeat France in six to eight weeks and then defeat Russia within a few months at most.

Those predictions, however, proved wholly incorrect. As the German offensives stalled outside Paris and in the Russian steppes and forests, the European nations settled into their long-term and bloody defensive grappling on the western and eastern fronts. Food quickly assumed an unexpected strategic importance, particularly for the Allied nations. Great Britain already imported considerable amounts of food even before the war began, and the German invasion of northern France deprived France of much of the land that normally supplied much of its grains. Within a few months of the outbreak of the war, the contesting

European nations, Allied and Central Powers alike, looked to the United States as a source of financial support and war material. Chief among their raw material needs was food, especially wheat, for their armies and legions of civilian industrial workers who poured into the war-production factories. The Allies bought and begged for American material, and, at the same time, neutral nations that imported American products also demanded their regular shipments. Although the U.S. government struggled to maintain its official neutrality, impartiality was all but impossible. Great Britain, with its large merchant fleet, was in a far better position than any of the Central Powers to exploit America's production capabilities. It used its formidable navy to protect that shipping and choke off German efforts to import American supplies. That reality, combined with historical and cultural ties between the United States and the Allied nations, quickly transformed America into the quartermaster for the Allies during the conflict.

American farmers were well positioned to meet the surging European demand. The national land policy in the years before the county joined the war effort, including the Stock Grazing Homestead Act of 1916, disposed of public domain lands to farmers and ranchers at reasonable cost and made vast areas of land available for increased production, especially of cereals. Cattle ranchers, too, had sold much arable land to farmers, who converted it from grazing to grain. Between 1914 and 1919 wheat cultivation land, for example, increased by 27 million acres, much of it on land located in the northern plains. The war coincided with one of the regular, cyclical wet periods in North America that ensured sufficient rain across much of the continent. New technologies and machinery were steadily improving farm efficiency. Surging European consumption translated into higher prices, increased national prosperity, and greater world market share for Americans. Between 1914 and 1917, corn prices increased, on average, from $0.71 to $1.45 per bushel, and wheat increased from $0.97 to $2.04. In one spectacular wartime price hike, the price of wheat in Chicago in June 1916 was about $1.06 per bushel, but by May 1917 the price had gone to $3.40— one estimate of total wheat profits showed a total increase from $56.7 million to $642.8 million between 1913 and 1917. Reduced national unemployment and the addition of regular wages paid by war industries also increased domestic demand for goods. Farm values increased as farmers competed with each other for cropland. Nationally, the price of farmland increased about 60 percent between 1916 and 1920, and the value of the most productive lands nearly doubled. Enhanced land values also increased the availability of credit for farm expansion and improvements. Although some of the increased prices for land and crops were the result of general wartime inflation, the gains in most cases were genuine. Such profits, reaped directly or indirectly at the expense of clashing European nations, helped define the period as the "Golden Age" of American agriculture.[32]

Then, in April 1917, the United States declared war on the Central Powers. American entry into the conflict required the agricultural sector to produce even more to meet the continuing needs of the Allies plus the new requirements of national war mobilization. Americans in the service and at home would need to be fed, and military and industrial manpower demands meant that there would be fewer young men to work the land. Farmers had the potential to increase their output to meet domestic wartime demand for food and fiber, but, like American society as a whole, they had done little to prepare for direct participation in the war during the years of American neutrality. The federal government was especially concerned about the ability of farmers to meet the increased demand for food and fiber that would result from the national war effort. Most Americans had believed that, even if the nation officially entered into the conflict, it would simply formalize the nation's de facto status as the chief banker and supplier for the Allied cause—few Americans during the period of neutrality expected that they would actually take up arms. Many Americans, especially in the heartland, were indifferent to fighting a war in Europe. Midwestern farms and cities were also home to a large number of first- and second-generation immigrants who retained close ties to nations of the Central Powers. During the early spring of 1917, with the threat of war looming, the federal government was already working to convince the agricultural sector to ready itself for possible American involvement in the conflict. "The best of the earth's people today," wrote the Secretary of Agriculture David F. Houston, "are in a death grapple. . . . In the coming conflict the farms of America must not only feed our people but much of Europe. When the struggle is over starving Europe will need food as the world never required food in the past. Mr. Farmer . . . remember your task this year is the heaviest ever expected from you. Your good wife and the labor of all the children are needed to make the farm produce as never before."[33]

Such patriotic appeals were unsuccessful, however, and farmers largely ignored Washington's war preparedness rhetoric. By the time the United States declared war, projected agricultural production for 1917 was about the same as the year before. The fact that the nation entered the war in the spring meant that many farmers had already committed their fields for the season. Crops were planted in many areas and could not be changed, and additional planting was also nearly impossible. Agricultural labor shortages occasioned by enlistment, Selective Service levies, and the attraction of wartime industrial jobs, distribution networks that were suddenly swamped by the competing demands of war industry and food production, and general rural inefficiency hampered the nation's conversion from a neutral bystander to a combatant. Washington's concern about the ability of the farmers to make the transition to a war footing led to measures that increased federal involvement in agriculture. During the opening months of America's participation, Congress passed sweeping legislation, including the Food and Fuel Control Act, that gave the federal government power to manage the nation's agriculture sector. New laws empowered the USDA to allocate seed, chemicals, and other agricultural goods that might be in high demand. The U.S. Food Administration, headed by Herbert Hoover, acted with a nearly free hand to control the national allocation of food. It wielded power over food processors and distributors and dictated what and how much they could sell and what prices to charge. It regulated exporters by telling them what they could sell, to whom, and how much they could charge. It worked to keep consumer prices at reasonable levels while still providing sufficient profits to farmers. Through the use of elaborate propaganda, the Food Administration stressed food conservation. It persuaded Americans to can produce they raised in "victory gardens" and observe "meatless" and "wheatless" days each week. It convinced Americans to substitute nonessential items for strategic ones as, for example, in its campaign to promote honey and sorghum syrup as a replacement for refined sugar. Administration propaganda even altered fashion tastes—shorter skirts saved cotton, and the slimmer female figure replaced the more matronly one that had long been Americans' ideal female form.[34]

DeKalb County farmers, like most in the midwestern corn and wheat regions, prospered during the period of neutrality. In the months before the war broke out in August, 1914, for example, DeKalb corn sold at between $0.70 and $0.80 per bushel, about the same or slightly higher than it had sold at its peak in 1913. After August 1914 and until the middle of 1916, European demand pushed corn prices up 10–20 percent, or between $0.80 and $0.90 per bushel. Beginning at harvest time 1916, however, prices shot up to $1.10 per bushel and continued to escalate at a remarkable pace. By early 1917, rumors of U.S. participation in the war and feverish speculation had driven corn prices to $1.70 per bushel. Hog prices followed similar trends.[35] The soaring prices were a clear indication that domestic production was lagging behind wartime demand, and, like most of the nation's agricultural sector, DeKalb farmers were not fully prepared to produce for national mobilization when America declared war.

In November 1917, Secretary of Agriculture Houston reflected that at the start of America's direct participation in the war seven months earlier, the country "was not fully prepared for war in any respect; but it was fortunately circumstanced in the character of its agricultural organization and the number and efficiency of its expert agencies."[36] The nation's leading agricultural agency was the USDA, which relied upon county extension services and agricultural advisors, like Eckhardt, to play a leading role in enforcing the broad range of wartime federal policies at the local level. Necessity, rather than experience, forced these men to assume responsibilities of organization and control that were unrelated to their original agricultural training and mission. Their primary task was to work with farmers to boost agricultural output to meet the wartime demand, but they scrambled to find and allocate seed, machinery, chemicals, and capital necessary for wartime production in their counties. As supplies of farm labor shrank, they searched for workers. They advertised in local newspapers, rounded up transients, encouraged town folk to help out during harvests, and dissuaded laborers from leaving the fields to take war-industry jobs.

In addition to helping with the economic mobilization of rural America, agents took on important new duties for social and political mobilization as well. Agents often advised draft boards about deferments for essential farm labor. They also did their best to comfort farm families that their drafted sons would return home once the world was again at peace. Some sons, of course, never returned, and advisors dealt with that, too. Many agents assisted with the Committee on Public Information, America's wartime propaganda agency, to overcome rural opposition or indifference to the war. As community leaders, they were expected to participate in the Liberty Loan campaigns and convince farmers, many of whom were reluctant investors and suspicious of bonds of any kind, to buy Liberty Bonds.[37]

Eckhardt, already busy with local farm problems and involved in agricultural politics at the local and state levels, found that he, too, was called upon to perform new wartime jobs. In late April 1917, just days after the United States declared war, Illinois state agriculturalists met in

Urbana to lay plans to mobilize farmers for the new challenges. They recommended, among other changes, that the nation establish a sort of draft to assure a suitable number of agricultural laborers, that farmers increase livestock production and combat diseases, that farm families grow garden plots, and that local newspapers stress to farmers that "[y]ou cannot raise a crop next winter. You must do it now. This is war." They were also sensitive to the dilemma facing young rural men who remained at home when so many entered the service. "It will take as much grit," reported one local newspaper, "for the young man full of red blood and true to the great traditions of our past to stay at home and care for this end of the work as it will go into the trenches. The man who does not shoulder one or the other of these tasks with all his might has no right to live under the stars and stripes."[38]

Securing adequate farm labor was as much a concern in DeKalb as anywhere in the nation. To manage the problem, the Soil Improvement Association served as a sort of labor exchange, and the *DeKalb County Farmer* regularly ran recruiting advertisements. "We have supplied many men with work," read one such advertisement, "and farmers with help. We will do our best to take care of your needs—get you a job or furnish you help." The Soil Improvement Association also assisted the United States Boys' Working Reserve in placing laborers on DeKalb-area farms. Eckhardt even suggested that the labor problem was so acute locally that "strong girls help when necessity requires." Like many of his contemporaries, Eckhardt was an active advisor to the local district draft board and helped several local families with petitions to the board for exemptions or discharges from military service so that farmers could continue to produce for the war effort. Eventually, the board granted seventy such petitions. He was immediately selected as chairman of the DeKalb County Food, Fuel, and Conservation Committee. The Soil Improvement Association also supported bond drives and war loan programs, and, perhaps more important, Eckhardt and Association leaders made patriotic appeals designed to convince local farmers that their work was vital to victory. "It is the call to arms and the appeal for bread that your nation now asks your help," wrote Eckhardt. "The citizen who believes in the kaiser and accepts the protection of this government had better go to his kaiser or hell for he and his will be hated and looked upon with contempt. . . . The farmer is asked to do his best and nothing short of this effort gives him a right to this free land and its abundant rewards. . . . Save yourself by planning your work and make every acre and every animal count as never before."[39]

Eckhardt's most important contribution to the war effort, however, was his ongoing effort to increase farm output. The key to increasing production was locating seed. Despite the wartime seed shortages, the Soil Improvement Association and the new DeKalb County Agricultural Association put DeKalb County in a better position than many Illinois counties to acquire quality seed because of their experience with seed brokers and area seed farmers. Eckhardt encouraged farmers to plant winter wheat in the fall of 1917, and the Association secured 1,300 bushels of Turkey red winter wheat seed for area farmers—it was the first time the Association handled winter wheat seed. The Association also purchased 11,000 bushels of Marquis wheat for spring 1918 planting. Marquis wheat produced about 20 percent more than the wheat commonly grown before the war. The widespread use of a single variety of wheat enabled local elevators to pay a higher price for wheat than if area farmers grew several different varieties.

The Association had little trouble securing wheat, but other seed was in short supply. Clover, alfalfa, and timothy were scarce, and prices were high. Corn seed, however, was in especially short supply throughout northern Illinois. Most farmers traditionally relied on their own harvest to provide seed for the following season, and Eckhardt regularly encouraged farmers to be careful about selecting open-pollinated corn for use as next year's seed. At the time of America's entrance into the war, there were no commercial seed corn growers in northern Illinois, and few commercial seed houses could provide high-quality seed stock acclimatized to DeKalb's growing season. The corn seed problem came to a head, however, in September 1917, when an early killing freeze destroyed much of the seed corn crop in the fields and left many farmers without seed corn to plant the next year. At the end of 1917, there was not enough corn seed in Illinois commercial seed houses to plant five corn belt counties. It would have been a difficult setback under any circumstances, but during wartime it was an emergency that demanded swift action. The Association advertised in the *DeKalb County Farmer* that it was buying as much as 1,000 bushels of seed corn selected from local fields before the frost and that it was willing to take "any variety that has done well a few years and is fairly pure."[40]

Corn belt farm advisors and their support organizations called a meeting to determine the best course of action. The representatives agreed to create a state seed corn "administration" empowered to locate and buy seed corn in volume and distribute it to midwestern farmers. Because of his experience in bulk seed buying, the group elected Eckhardt as State Seed Corn Administrator to undertake the work. The Soil Improvement Association agreed to lend their advisor to the Illinois State Council of Defense, located in Chicago, beginning January 15, 1918. Charles L. Gunn, the Association's assistant demonstration agent, took over Eckhardt's duties in DeKalb. Seventeen Chicago banks agreed to immediately loan the administration $1.25 million to buy seed corn. Eckhardt used his professional contacts and influence to help the administration buy two-year-old seed corn and early maturing seed that had escaped the freeze, test it, and distribute it to farmers in Illinois, Indiana, and Iowa. He also enlisted county agents, bankers, and farmers to report the sale of any inferior or untested corn seed. At the end of his term Eckhardt reflected on his experience. "The entire county agent force of Illinois was organized into one machine. Where seed corn existed the agent made available the seed, where there was a shortage he sold seed. Nearly 125,000 bushels of seed corn was bought, sold, and delivered through this organization. It involved a handling of more than $1,600,000 with no record of bad seed furnished or funds misused."[41]

When the war ended in November 1918, American farmers were in a precarious position despite their wartime profits. The war years were a time of artificially high demand, surging wealth, and high inflation, but the opportunities to improve farms, purchase machinery, and enhance living standards were limited by the demands and shortages occasioned by the war. With the war over and farmers awash in cash and other assets, farmers invested their wealth in new land, triggering an agricultural land boom. By 1920, land prices nationally were about 70 percent higher than they had been during the period 1912–1914, with about 40 percent of that increase occurring just between 1918 and 1920. In some of the corn belt areas, land doubled in price by 1920. In DeKalb, rising property values are reflected in property taxes collected on farm real estate, which increased from about $267,000 in 1913 to $330,000 in 1918, $435,000 in 1919, and $530,000 in 1920.[42] Although much of the high prices were speculation-fueled paper values, farmers were spending their cash and incurring mortgages to meet the inflated prices. Net farm debts increased 250 percent between 1914 and 1920. Speculation in purebred animals and investment schemes of all kinds also converted real farm cash and assets into fragile paper profits.

Many factors fueled such foolhardy optimism. Domestic purchasing power, postponed by the war, exploded after the armistice as Americans eagerly sought to buy what had been unavailable during the conflict. The European nations, struggling to prevent mass starvation and rebuild their war-shattered economies, continued to purchase American crops after the war ended, keeping the domestic wartime agricultural boom alive. Credit and loan policies followed by the Federal Reserve Board and the treasury department masked underlying instabilities of the agricultural market and postponed the inevitable postwar contraction for nearly a year and a half. Perhaps most important, however, the agricultural sector simply had too little experience with economic conditions caused by the war and the postwar boom. In late winter of 1918, prices slumped, as many farmers expected, but then quickly recovered. By May 1920, agricultural prices were nearly 175 percent higher than they had been in 1913. Corn prices in DeKalb fell from wartime peaks to about $1.30 per bushel in late 1918, climbed in mid-1919 to around $2.10, fell again to around $1.60 in late 1919, and rebounded to more than $2.00 in mid-1920. Consumer food prices nearly doubled 1913 levels, clothing and cloth prices were about 250 percent higher, and home furnishings were about 270 percent higher. Many rural Americans believed that the war had ushered in a new era of permanent prosperity for the agricultural sector and that higher price levels for their commodities were no longer exceptional. These assumptions inflated land and stock prices, fueled mortgage borrowing, and encouraged financial risks and carelessness that would have been unthinkable before the war.[43]

The short postwar boom reached its peak during the autumn of 1920. Then the inevitable postwar contraction finally occurred. Tremors of the economic crisis were already noticeable in DeKalb County as early as December 1920, when farm auctioneers reported that auctions were bringing in the lowest prices than anytime in the preceding ten years. Thereafter, agricultural prices plunged at a rate that was as spectacular as the 1916–1918 increases had been. Corn in DeKalb in mid-1921 sold as low as $0.40 per bushel and mid-1922 at $0.60 per bushel. Because the high agricultural prices had been the result of a series of artificial conditions related to the war, peace brought with it a return to market conditions that prevailed before the conflict. Farmers had come to rely on the new prosperity. Consequently, despite the fact that the agricultural sector had just enjoyed the most prosperous period in its history, when the farm economy crashed, American farmers found themselves poorer, and their farms more worn out and more heavily mortgaged than they had been before the war began.[44]

By the spring of 1921, American agriculture was in serious economic crisis. The return to more typical supply-and-demand conditions after the war put a downward pressure on farm prices. Farmer purchasing power in relation to nonagricultural products had fallen nearly 65 percent compared to prewar levels. Countless farmers were burdened with large mortgage debts for land and equipment bought at inflated prices during boom times but now worth only a fraction of the purchase prices. Rural banks were papered with farmers' notes that could no longer be paid, and the resulting epidemic of rural bank failures swept away investors' savings while leaving heavy debts. European demand waned as those nations found it increasingly difficult to pay for American goods and their own agricultural sectors slowly recovered to meet their domestic demands. Farm income from exports declined nearly 40 percent between 1919 and 1929. Domestic purchases also declined. In a society that glorified a new slim and fit image of beauty, Americans ate more fruit, vegetables, and diary, but consumed fewer starches, fats, and meats. Shorter skirts and synthetic fibers, especially rayon, cut demand for natural fibers. Moreover, farm prices were highly inelastic. Whereas falling prices for durable goods, such as automobiles and radios and other farm family necessities, induced consumers to buy more of those items, the same was not true of farm prices—no matter how low commodity prices fell, Americans could not consume more farm output than they did already. Consequently, costs for goods and services remained high in relation to farm income.

The rapid increase in agricultural prices and inflation occasioned by the war make the declines of the 1920s appear greater than they were, and some scholars suggest that the agricultural depression was rooted more in statistics than reality.[45] American farmers, after all, found their prices falling to about the level of the prewar period, a time when farm prices were relatively high historically. Those who suffered most, they argue, had simply overextended themselves during the boom years. Yet, despite statistics, for many farm life was harder during the 1920s than during the war years. Rural real income was less than during earlier periods, primarily because prices for nonfarm goods and services did not fall as fast as farm incomes did. Twentieth-century farm families were also less prepared than earlier generations to weather the downturn. For years, after all, urban philanthropists and thinkers, agricultural suppliers, railroad spokesmen, mail-order merchants, the media, land-grant college professors, extension service and farm agents, USDA policy makers, and a host of other farm reform voices had told rural Americans that they should abandon their old ways of doing business and modernize their operations with new, and expensive, technologies. They encouraged farmers to compare their living standards with those of their urban contemporaries, not each other, and to demand a similar quality of life. Farmers during the 1920s, therefore, were generally less self-sufficient, bound to complex new technologies, dependant upon outsiders to provide their material goods, accustomed to higher living standards, and committed to paying higher taxes for rural public improvements such as roads and schools. Most American farmers found it difficult, if not impossible, to return to simpler forms of production and living. For the half a million or more American farmers forced off their farms by bankruptcy, rural bankers who lost their banks, depositors who saw their savings vanish, and rural merchants whose businesses failed, the hardships of the period were intense. In some places in the country, the very structure of the local economy was threatened. In parts of Nebraska and the Dakotas, corn prices were so low that farmers burned it for fuel. In many areas of the American countryside, barter, which had always been a part of rural life, briefly supplanted cash exchange. By the winter of 1921, the economic collapse was complete, and the difficult period of readjusting the farm economy to a more traditional, and harsh, postwar supply-and-demand footing had begun.[46]

When Eckhardt returned to DeKalb County from his job as wartime State Seed Corn Administrator, the immediate postwar agricultural boom was in full swing and few anticipated that the prosperity would end. Membership in the Soil Improvement Association had increased nearly five-fold—from 400 in 1918 to 2,024 a year later.[47] Demand for services from the farm advisor, the Soil Improvement Association, and the Agricultural Association was so great, and Eckhardt was so deeply involved in state agricultural matters, that in June 1919 the Soil Improvement Association hired Thomas H. Roberts as its new assistant farm advisor. The Association also raised dues from $2.50 to $5.00 to cover the costs of their expanding services and the new assistant. Roberts, born in Waterman, DeKalb County, attended Iowa State College and Northwestern University and learned the practical business of agriculture on his family farm. A few months later, Eckhardt was injured in a household accident and resigned his position as farm advisor, effective April 30, 1920. A short time later, he took a job as director of grain marketing for the Illinois Agricultural Association. In truth, after eight years as farm advisor, Eckhardt was ready to change jobs. His work in agricultural matters at the state level increasingly drew his attention away from his local responsibilities. In particular, Eckhardt's experience during the war and his interest in cooperative action suggested to him that the best financial option for midwestern farmers was to market their grain through a uniform, well-organized marketing system in the large producing states. Only then could farmers have sufficient economic power to influence grain prices. The marketing program would be managed by trained, professional agents. These large cooperatives could also buy farm necessities in volume and pass the savings on to members. In July, the new American Farm Bureau Federation, the national umbrella organization for farm bureaus throughout the nation, appointed Eckhardt to the "Committee of Seventeen."[48] The committee was charged with developing a plan by which all the organizations of grain producers could operate through centralized grain exchanges. The new cooperative association would support and organize local cooperatives, supply warehouses and terminals, finance grain storage, and export pooled grain. Although Eckhardt remained for a few months as a part-time DeKalb County advisor and as a $1 per year "manager" of the Soil Improvement Association, the Illinois Agricultural Association wanted to avoid any conflicts of interest and requested that he end all association with the DeKalb Soil Improvement Association and the DeKalb Agricultural Association. By July, 1920, the Soil Improvement Association needed a new advisor.[49]

The Soil Improvement Association turned to Thomas Roberts to take over the job of advisor. A short time later, in October, 1920, it also hired Orr M. Allyn, a stockman from Illinois and Montana, as the new assistant advisor. These two men assumed their new duties during the most difficult economic times in American agriculture in a generation. Moreover, in DeKalb County the role of the Soil Improvement Association and its county advisors was changing dramatically. Although the Association remained committed to its original education mission and enjoyed continued success in teaching and encouraging new agricultural techniques, economic conditions demanded new ideas and greater involvement in the community. The Association could no longer focus simply on soil preservation, livestock raising, efficient production, and seed marketing, but also needed to offer innovative new programs to county residents. "[W]e believe," noted Roberts in 1920, "the educational part of our work is not one of the chief problems today."[50] It became increasingly evident that improving the economic health of Association members meant devoting more time to developing cooperative marketing, collective purchasing, and legislative activities.

Beginning in the brief postwar boom period, Eckhardt and the Soil Improvement Association were broadening the range of services they offered DeKalb farmers. When Roberts took over the job, he continued their work to expand Association services and integrate the Soil Improvement Association even more closely with the local community. During the summer of 1919, for example, Eckhardt had proposed that the Soil Improvement Association work with the Agricultural Association to build a grain mill in DeKalb, but the plan moved ahead slowly. When Roberts took over, he and the Association successfully sought community support for the plan. It was a sizable undertaking, but by eliminating as many as eight middlemen and charging farmers and local citizens only the cost of milling and retailing, Roberts estimated the new mill might save DeKalb County citizens $250,000 annually—a year later, in 1921, the completed mill produced more than 440 tons of flour and cut processing costs by 20 percent. The Soil Improvement Association and the Agricultural Association arranged the purchase of 31 railcar loads of Idaho apples and 110 cars of potatoes for residents of DeKalb and neighboring counties. Roberts, who was a capable businessman, assisted 325 farmers with their income-tax returns during the spring of 1920.[51] The Soil Improvement Association worked closely with local merchants and community organizations—including the Kiwanis, Odd Fellows, Rotary, Chamber of Commerce, and Commercial Club—in support of a range of civic projects. In contrast to Eckhardt, Roberts was agreeable to the idea of working with children and arranged for a pig club, in which high school students worked together to improve county swine herds.[52]

Livestock and grain marketing, however, quickly moved to the forefront of Soil Improvement Association priorities in the climate of economic uncertainty. Roberts noted in late 1920 that "the marketing problem is the most serious problem facing DeKalb County farmers today." To solve the problem of high yields and low returns, the Association helped to organize a series of cooperative ventures. Although the development of cooperatives was left to farmers themselves, the Soil Improvement Association sponsored speakers and discussions about cooperative action and publicized organizing events. Roberts and Association officers offered advice. During the early 1920s, farmers organized a cooperative elevator, located in Sycamore, two livestock shipping associations, and a county wool-marketing pool. Some of the cooperatives remained modest affairs with a small membership, but others grew considerably. The livestock shipping association based in Clinton Township had ninety-nine members and shipped eighty-three cars of livestock during its first year. The wool pool shipped nearly 50,000 pounds of wool in 1920 and more than 39,000 pounds a year later.[53]

In addition to his work in enhancing marketing opportunities, Roberts believed that the chronic credit shortage was the other pressing agricultural problem facing DeKalb County farmers. "Money and credit to finance feeding of livestock, the payment of farm mortgages which are maturing and the financing of farmers so that they may market their grains in an ordinary manner," wrote Roberts, "have been the crying needs of our community for the last year." Because of fluctuating commodity prices, interest rates, and property values, banks, mortgage firms, and insurance companies traditionally made loans for less than five years; most loans were for three years and many were for one year only. These loans could not be easily paid on time under the best of circumstances, and neither the lenders nor the borrowers expected them to be. Usually, when a loan came due the parties simply renegotiated the loan rate and duration, and the process was repeated. Such a revolving loan system provided lenders with a method for holding an equity interest in farmland while retaining the freedom to change the loan terms quickly to reflect market conditions.

Such a credit system placed a heavy burden on the farmer-borrower. He was never secure, because loan renewals were so frequent and new applications were subject to denial. A rejection would be disastrous if it came at a time when financial conditions were poor and commodity and land prices were down. Moreover, loan-servicing fees, such as title insurance, commissions, and survey work, were costs that needed to be paid by the borrower at the time of renewal. Farmers demanded ways to extend mortgages and more efficient means of repaying loans, yet small, rural commercial banks were often ill-equipped to undertake the risk of a new loan procedure and large institutional mortgage holders had little incentive to change the system. "There is a much needed form of credit which is not now available," wrote Roberts, "to take care of farm loans running from 12 to 30 months before maturing. Our present banking laws have been devised solely with the view of protecting the depositor and without regard for the borrower, especially the borrower who had to have more than six months time. This is not a criticism of our banks, but of economic conditions which have brought about somewhat of a hardship on farmers." To make matters worse, the drop in commodity prices at the end of the war coincided with a steep decline in land prices. Farmers who had mortgaged their land during the flush times found that they were unable to pay off their loans or persuade their creditors to renegotiate their loans. Many farmers teetered on the edge of bankruptcy or risked slipping into tenancy.[54]

One response to this problem was the federal Farm Loan Board, a system of cooperative federal land banks and joint stock land banks organized by private capital for private profit, but operating under a set of credit-easing federal guidelines. When the Farm Loan Bank opened in DeKalb, Roberts was appointed acting secretary-treasurer. Eventually, the Farm Loan Bank received $250,000 in loan applications from area farmers and loaned more than $216,000 to them in 1920—by the end of 1923, the Farm Loan Bank had loaned in excess of $936,000 to seventy-six DeKalb County farmers. The Soil Improvement Association assisted farmers in making local applications to the Joint Stock Farm Loan Bank totaling more than $200,000. Although Roberts, like many of his constituents, was initially skeptical about the government's role in such a local matter as farm finance, he nevertheless realized the important security it provided and eagerly assisted farmers with securing these federally regulated loans. In addition to making credit more available, the federally regulated loans offered farmers some tax relief, and the competition with local banks lowered interest rates, commissions, and terms across the county. "We feel that the Farm Loan Act has been the most helpful legislation that was ever designed to help the farmer. . . . We feel that every farmer who has a farm mortgage has been benefitted by this Act regardless of whether he took a loan out thru the Land Bank or not, as their competition has lowered interest and commission rates on farm mortgages." Roberts was also instrumental in establishing the DeKalb County National Farm Loan Association, with the Soil Improvement Association as its headquarters. The local Farm Loan Association loaned money directly to farmers for less fixed cost and at terms that were usually more favorable than the commercial mortgage lenders.[55]

The Soil Improvement Association's emphasis on land financing during troubled economic times is understandable. On one level, it reflected the prevailing American assumptions about the inherent importance of farmers owning the land they worked. Farmers burdened by large mortgages and living on the edge of bankruptcy were not the sort of successful farmers who formed a solid foundation for community prosperity. Moreover, in comparison to other parts of the country, DeKalb County had few absentee landlords. When credit conditions were favorable and the mortgage burdens reduced, financially secure landowners were less reliant on rent income and more likely to sell land they were unable to farm themselves to their tenants. "It is not our purpose," Eckhardt wrote, "neither is it desirable to have a great many changes take place at once. If we can in 25 years have 90 percent of our farms owned by the men operating them, our task will be well done."[56] Perhaps more important than land security, however, was the internal dynamics of the Soil Improvement Association itself. Bankers were instrumental in founding the Association, and county bankers remained influential Association members—Roberts himself was the member of a Clinton Township banking family. During the 1920s, many local banks were burdened with high-risk loans and a shortage of capital. Many of the bankers supported the Association's financial efforts because they, and their banks, were landowners themselves who saw the Association as a means of establishing farm security throughout the county. County farmers who refinanced through the Federal Land Bank released bankers from potentially ruinous mortgage exposure and returned capital to the banks.

All of these new services were important, but the Soil Improvement Association enjoyed broad support because it never lost track of its original mission. Roberts regularly made visits and prescribed soil improvement techniques and had a busy office schedule. In 1921, for example, Roberts made nearly 400 farm calls, had 5,700 office conferences, wrote 7,500 letters, wrote 50 agricultural articles, and mailed 65,000 copies of 19 circular letters. Membership in 1920 and 1921 slipped slightly to a total of 1,875 members, primarily because of the difficult economic conditions. The seed business expanded, and the Agricultural Association, which by 1921 had expanded to $100,000 in capital stock, shipped 410 tons of limestone and 200 tons of phosphate, in addition to providing members with high-grade seed. Roberts's vision for DeKalb County farmers was simple: "Efficiency in production of farm products with special emphasis laid on so arranging our farming business that we can operate with the lowest cost, possibly by employing less farm help and using more legumes with less grain crops, will do much to solve the tremendous surplus which is now flooding the market and also increase the farmer's returns for his efforts. Taking all in, although the present is dark and gloomy for the farmer, there is one bright spot in the fact that the farmer has got to have buying power before industry can operate at a profit."[57]

Despite this hopeful assessment, in 1922 the agricultural recession meant that even the generally successful DeKalb farmers found it difficult to make a profit. Roberts noted, for example, that he "did not find a single instance where a man, interested purely in farming, had to pay an income tax." Just three years earlier, Roberts had devoted considerable time to helping 325 farmers calculate their income taxes and reported that 259 sent in returns. Soil Improvement Association membership declined to 1,001, down more than 800 members from the year before. Farm wages were low. The Association helped form six new livestock shipping associations in the county to try to reduce livestock transportation costs even further. The county wool pool, which had started out so strongly two years earlier, shipped only about 6,000 pounds of wool. Roberts worked with area farmers to substitute winter wheat for more traditional oats, a change that he predicted would more than double the income per acre. Trying to put the best face on the situation, Roberts wrote, "All things considered, we have gone through a year filled with a lot of grief. This has necessitated double the effort on our part to put over our projects, but we feel that considering financial conditions we have gone through this period of readjustment in very good shape and are probably the stronger for having had these experiences."[58]

In the midst of the difficulties, however, there was also time to celebrate. The farm advisor movement in DeKalb County and throughout Illinois survived its early years, the challenges of the war, and the dismal postwar economy. Ten years after Eckhardt began working as DeKalb's agricultural advisor, the Soil Improvement Association, which by that time was already one of the most successful and prominent agricultural associations in the state, hosted a festival to commemorate the tenth anniversary of the farm advisor movement on June 29 and 30. It was a grand affair endorsed by the Illinois Agricultural Association, the Illinois Association of Farm Advisors, and the University of Illinois. Henry Parke was selected general chairman of the event.[59]

The American Farm Bureau Federation declared with little modesty that the festival was the "greatest spectacle of its kind ever attempted." Regional railroads granted a fare discount to those who could not drive. At one point, organizers almost persuaded President Warren G. Harding to attend, but he eventually declined. It was also a great opportunity to showcase DeKalb County. "Everyone who has any pride in DeKalb County," wrote Roberts, "should take a part in seeing to it that these thousands of people go home with the best possible opinion of DeKalb County."

For two days, between 20,000 and 30,000 visitors from Illinois and neighboring states celebrated the farm bureau movement and the achievements of modern farming. Public speeches, lectures and demonstrations by university professors, farm journal editors, and even Minnesota Governor Jacob Preus highlighted the history and future of American agriculture. The DeKalb Military Band, the Barb City Quartet, and the Boys' Band of Savanna, Illinois, entertained audiences with patriotic songs and popular favorites. Parade floats from fifty-five mid-western counties drifted up DeKalb's main street highlighting such agricultural topics as soil testing, sweet clover, poultry culling, chinch bugs, limestone, and swine sanitation. Not surprisingly, DeKalb County's float theme was "Pure Seed for DeKalb County." "Forward! Farm Bureau," a four-episode play written for the event, dramatized the birth of the farm advisor movement and its importance for the future of rural communities. The drama featured 4,000 actors from fifteen counties and was directed by Miss Nina B. Lamkin, of New York, a professional outdoor pageant director.[60]

For the county farm bureaus and their members who participated in the celebration, it was an opportunity to reflect on their success. In just ten years, membership in farm bureaus had grown to around 100,000 farmers statewide. The event also reflected the underlying social attitudes and expectations held by many in rural Illinois. In his keynote speech, Governor Preus articulated what he believed were the most important ones and in so doing clearly distinguished the essentially conservative farm advisor movement from the more left-leaning farmer organizations that were attracting members during the difficult economic times. After equating the Non Partisan League with the Bolsheviks, he praised the farmers of DeKalb County. "[Y]our influence has always been for the best interests, not only of its members," he said, "but for the whole community. Perhaps one reason is that your organization is composed entirely of real farmers and controlled by them. Another reason is that you have remained true to the principles of our constitution and the ideals of the founders of our government, and you have not been led astray by those sinister influences which from time to time seek to destroy the American ideals. . . . Your federation has proceeded along safe and sane lines and it has not attempted to overthrow our American ideals of life, liberty, and private property but is working for the continuation and the perfection of those institutions which have brought the American republic to its present state of power and influence in the struggle for the welfare of mankind."[61] The farm advisor movement had always been an organization that worked to help farmers accommodate themselves to the modern American economic structure without abandoning national ideals that cast farmers as the noble, Jeffersonian embodiment of traditional American values. From its beginning, it was concerned more with fostering financial success than on social reform, and, according to Preus and his enthusiastic audience, the expression of those ideals was one of the movement's greatest accomplishments.[62]

Despite the festivities, economic problems continued to haunt DeKalb County during the 1920s. At the meetings of the Soil Improvement Association, financial concerns and membership were regular items on the agendas. In early September 1922, Association spokesmen reminded members that farm purchasing power was at least 30 percent less than it had been during the war years. They exhorted local businessmen to help raise the county out of "the chaos into which it is sinking." "Agriculture," they wrote, "is the basis of all industry and until it can be made profitable we need not expect any prosperity." As if to underscore the uncertainty of the era, as part of its recruiting effort to enroll 1,800 new members, the Association planned to show films of June's pageant and present a lecture entitled "Where Do We Go From Here?"[63]

Nevertheless, Roberts and the Association were determined to continue offering area farmers relevant information and practical assistance to help them remain afloat financially. The advisor's office sent thousands of pieces of mail, and Roberts made dozens of farm visits each year. The DeKalb Agricultural Association continued to import large quantities of high-quality seed and soil additives. By 1923, such activities made DeKalb County the fourth-ranking county in the state in average corn yield. Roberts and the Soil Improvement Association, however, realized that teaching farmers new techniques and providing them with quality seed was no longer sufficient to keep their members at the agricultural forefront. They remained committed to the idea that efficient marketing was the key to financial success for the county's producers and focused ever-increasing attention on helping farmers work collectively.

DeKalb County had several livestock and dairy operations. In fact, by the early 1920s the county's agricultural economy was based more on livestock than grain.[64] Consequently, Roberts and the Soil Improvement Association devoted considerable attention to livestock, with special emphasis on sanitation and the eradication of infectious disease, especially tuberculosis. During 1923, the Soil Improvement Association helped organize still more cooperative livestock-marketing associations, making a total of ten operating in the county, and assisted in negotiating favorable shipping rates and resolving trucking disputes. A year later, in 1924, the livestock shipping associations shipped more than 910 carloads of livestock out of the county. The county's new poultry association worked to improve the quality of locally produced eggs. The Soil Improvement Association also waged a three-year crusade to raise local interest and persuade the county board of supervisors to hire a county veterinarian, Dr. O. H. Lintner, as part of its livestock tuberculosis-eradication campaign.[65]

The Association advised the Sandwich and DeKalb Milk Producers Associations about collective marketing activities and worked hard to secure the necessary inspection certificates and loading facilities that would enable the dairymen to break into the Chicago fluid milk market. When new citywide public health standards threatened to cut the 350 DeKalb County dairymen out of the Chicago milk market in 1927, the Soil Improvement Association helped establish a countywide milk producers' association, worked with dairymen to meet the new standards, scoured the Chicago area for new customers, and negotiated a new, more lucrative milk contract. The DeKalb Agricultural Association leased a dairy building in Waterman, Illinois, remodeled it to comply with the Chicago Board of Health regulations, and served between 300 and 400 area dairy farms.[66]

As demand for services from the Soil Improvement Association grew and its offerings became more diverse, it undertook a series of new initiatives to complement its traditional education functions and more recent marketing initiatives. For example, Roberts, who had a business background, was an advocate of proper farm record keeping, and the Association provided free account books and sponsored classes on farm accounting. Roberts began offering advice on farm finances, assisting farmers in securing loans, and helping them negotiate lower interest on outstanding loans. The Association was instrumental in establishing a tractor school in Sycamore to train local men in engine maintenance. The Association ran an employment bureau for farm labor. During the slack season of 1923, Roberts worked to attract a canning company to the county, but he was ultimately unsuccessful. To combat Russian thistle, the Association members helped the county thistle commissioners enforce Illinois statutes prohibiting the spread of thistle. The Association supported the DeKalb County Farmers' Protective Association, a group dedicated to enforcing trespassing and hunting laws. Hunting regulations were especially important because increasing numbers of "careless hunters from Chicago" were accidentally killing local horses, sheep, hogs, and even chickens.[67]

Perhaps more important for the long-term success of county farmers, the Soil Improvement Association expanded its offerings to include more emphasis on educating rural children. The Association improved cooperation between itself and area schools. It assisted, for example, with a program in which high school students worked with local dairymen in hygiene testing and record keeping. The Soil Improvement Association invited area schools to display exemplary schoolwork at its functions. "This created a great deal of interest," noted Roberts, "on the part of the fathers and mothers at these meetings and resulted in a very good general attendance." The Association sponsored calf clubs and pig clubs that proved especially popular, and it helped to create a new dairy calf club and a poultry club. The Soil Improvement Association also provided much of the financing for the operations of these youth clubs.

In contrast to Eckhardt, who during the program's formative years had been generally opposed to working with children, the Association was so deeply involved in youth club programs by the summer of 1925 that it hired Raymond Nelson as assistant to the farm advisor. His duties were largely devoted to youth club work and assisting area vocational education teachers in the high schools. Members of these youth clubs competed successfully, a few even earning state-level recognition for their achievements. In 1925, the dairy judging team won the state championship and placed second nationally. Although a few girls participated in the stock clubs, the Association also embraced strictly girls' clubs for the first time. In cooperation with the state extension service the Association held training sessions for girls' club leaders and sponsored four local clothing clubs in 1926. Nelson secured eight Singer sewing machines for the clubs a year later. Nelson was also instrumental in reorganizing these county youth clubs under the umbrella of the newly created 4-H clubs in 1927.[68]

In addition to these new social programs, the Association was becoming more politically active and was quickly evolving into the leading collective voice of farmers in the county. In 1925, the Association filed a brief with the Illinois Commerce Commission in opposition to a proposed telephone rate increase for the county. It was one of the most persuasive voices in the rate defeat. Taxes also played a prominent role in the Association's political activities. When the Illinois Agricultural Association lobbied for the passage of a state income tax to help shift the cost burden of public services away from property taxes, for example, the DeKalb Soil Improvement Association agreed and worked to educate area farmers about the benefits of the new law and build local support for the measure.[69]

Property tax relief was an especially important issue for DeKalb-area farmers. During the opening decades of the twentieth century, commodity prices and farm incomes rose generally, enabling farmers to absorb the costs for new state and local government initiatives and services. Farm taxes continued to soar during the profitable war years at a rate faster than corporate taxes. When commodity prices collapsed in 1920, however, farmers were forced to pay taxes assessed at wartime values from recession-level incomes. Tax relief was on the minds of hard-pressed farmers nationwide early in the decade, and the Illinois Agricultural Association began its successful crusade for state tax reform in 1921. While the state real estate tax revisions saved Illinois farmers about $250,000, local taxes also needed to be reduced if DeKalb-area farmers were to realize any meaningful tax relief. During late 1924 and 1925, Roberts and Nelson drafted a "tax equalization" pamphlet, entitled "A Request for More Equitable Taxes in DeKalb County," that they distributed to DeKalb County Board of Review of Assessments and area residents.[70]

Their evidence demonstrated that farm values per acre in 1924 were nearly equal to what they had been in 1913, but property taxes for the county as a whole had risen more than 230 percent during that same time. The property tax schedule meant that farmers were paying a disproportionate share of county taxes. Moreover, DeKalb taxes were 36.6 percent higher on average than all other Illinois counties, including Cook, and nearly 45 percent higher than similar counties in other corn belt states. The tax assessor's forms listed and taxed every item of farm property, but merchandise in factories and stores was listed as an unitemized lot. Their statistical analysis of the county's seriously inequitable tax structure resulted in a 10 percent reduction in property taxes in 1925, and a further 10 percent reduction two years later. The Association's tax-relief campaign was no doubt helped by the composition of the county board—between 1924 and 1929, from one to four members of the county board were Association officers or township Association representatives.[71]

As if to underscore its rapid evolution, even the name of the Soil Improvement Association changed. For many years, county residents had casually referred to the Soil Improvement Association as the "Soil Improvers," occasionally as the "Soilers," but most often as simply "the Farm Bureau," a term that in other counties usually referred to county advisor programs organized pursuant to the Smith-Lever Act. Eckhardt used the term as early as 1919 in his *Annual Report of the County Farm Advisor*. Yet, although the Soil Improvement Association was organized prior to the creation of the federal extension service, the "Farm Bureau" nickname stuck, and even Roberts himself was using it regularly. Moreover, the Illinois Agricultural Association was encouraging counties to adopt the "Farm Bureau" name as part of its campaign to unify county advisor programs throughout the state. In DeKalb County, the activities of the Soil Improvement Association had expanded beyond its original mission of promoting soil conservation, and a name change seemed to be in order to reflect that new reality. Consequently, on April 28, 1926, the Soil Improvement Association board of directors voted to make the change official, and the Soil Improvement Association became the DeKalb County Farm Bureau.[72]

The Farm Bureau struggled through the decade of sluggish farm profits to maintain its financial health, but low membership put a strain on its budget at the very time it was expanding services. Despite these difficulties, the Farm Bureau during the late 1920s turned its attention to offering new financial services as part of its ongoing effort to provide innovative practical assistance to its members. One of the most important new programs was insurance. Securing hazard insurance for farm buildings was often difficult, expensive, and confusing. There were hundreds of small insurance companies, but farmers had to seek out the insurance agents, and policies offered poor coverage. Illinois state law also limited the amount of coverage that such small companies could issue on each risk. Consequently, most Illinois farmers were dangerously underinsured for their most important assets.

To address these problems, in late 1924 farm bureau leaders in several counties, including Henry Parke, urged the Illinois Agricultural Association to discuss with some of the state's largest mutual insurance companies the possibility of creating a new reinsurance company to protect farmers. Such a statewide reinsurance company would make casualty insurance more available and affordable because it spread the risk of loss over a larger population of farmers. In 1925, the Illinois Agricultural Association, under contract to promote the new insurance pool, began an aggressive campaign to sell insurance policies to Illinois farmers, with the Illinois Agricultural Association agreeing to cover the incidental operating costs. Throughout the summer of 1925, Illinois Agricultural Association representatives visited DeKalb and other Illinois counties and secured $751,000 in insurance applications. In late 1925, the Illinois legislature issued a new corporate charter to the Farmers Mutual Reinsurance Company, which quickly became one of the most important agricultural casualty insurance companies in the state.[73]

Parke, one of the founders of the Soil Improvement Association, was instrumental in persuading the Illinois Agricultural Association to create an auto insurance company for Bureau members. Statistics showed that farm vehicles were a much lower insurance risk than urban ones, but insurers charged farmers premiums based on country and city accident rates. Created in the spring of 1927, the Illinois Agricultural Mutual Insurance Company enabled farm bureaus to offer equitable auto rates to their members. In DeKalb County, the Farm Bureau quickly set up an auto insurance affiliate, the Country Mutual Insurance Company, that worked hard to sell the new Agricultural Mutual policies. The local response from Bureau members was favorable. "[M]ore interest seems to be shown in this project," wrote Roberts, "than in any other started by the Illinois Agricultural Association recently." By the end of the summer of 1927, the Bureau's Country Mutual Insurance Company had written 186 Illinois Agricultural Mutual auto policies—by 1928, it had written 231 policies. Bureau members were delighted with the accessible and affordable insurance and demanded a whole range of new insurance products. In 1929, the Farm Bureau expanded its insurance program to include another affiliated company, called the Country Life Insurance Company, to provide life insurance to members. Eventually Bureau members were able to buy a wide range of insurance products, including crop, fire, and casualty insurance, through Farm Bureau insurance affiliates.[74]

The other major new business venture for the Farm Bureau during the 1920s was oil and petroleum products. Before World War I, lightweight, powerful gasoline tractors were fast replacing draft animals and steam tractors on midwestern farms, a trend accelerated by the production demands and prosperity of the wartime economy. By the 1920s, petroleum products had became a farming necessity but were sometimes in short supply and of inconsistent quality. Moreover, farmers usually bought more oil and gasoline than their urban counterparts, and oil company delivery charges to farms in remote corners of the county increased the price. By the end of 1926, several farm bureaus statewide had formed cooperative companies to distribute petroleum products at a discount to their members. At the March 1927 meeting of the DeKalb County Farm Bureau, the members voted unanimously to have the DeKalb County Agricultural Association handle oil in addition to its more traditional seed and soil additives business. The Association and the Farm Bureau signed a contract agreeing that the Agricultural Association would rebate 50 percent of the net profits of the oil business to Farm Bureau members who purchased the oil.[75]

The DeKalb County Agricultural Association joined the newly incorporated Illinois Farm Supply Company, a corporation whose stockholders originally included nine county oil sales companies, including the Association. The Association's oil sales department, the DeKalb County Oil Company, leased land from the Great Northern Railroad, installed three 16,000-gallon storage tanks, and purchased two delivery trucks.

On the first day of operation, in August 1927, a secretary called county farmers to remind them of the new service, but it was hardly necessary. Andrew Shipman placed the first order that day, and driver Emil Rueff rushed to the Shipman farm to make the delivery. Business quickly boomed. By the end of the year, the DeKalb County Oil Company had four delivery trucks, and by the end of 1928 it had two storage facilities and six trucks. Each month, in addition to lubrication oil and grease, 431 regular customers bought two full railroad carloads of gasoline and two of kerosene. Profits rolled in, and member rebates flowed out—the first rebate checks to Farm Bureau members totaled $2,973. By the end of 1930, the Agricultural Association expanded service to include neighboring Boone, Ogle, and McHenry counties with twenty-one trucks.[76]

The tremendous changes that occurred during the 1920s, and the responses that the DeKalb County Farm Bureau crafted to help farmers succeed in the new economic and social climate, foreshadowed developments in agriculture for the remainder of the century and set the tone for DeKalb County Farm Bureau programs. The Farm Bureau was always most successful at developing programs that fulfilled specific needs of county farmers. In the decades that followed, farming would become increasingly mechanized and dependant upon scientific and technological innovation. The size of farms would increase while the number of people who made their living directly or indirectly from farming would decrease. The business of farming meant that farms were heavily capitalized and more commercial than ever before. Rural families became better educated. Few leaders of the Soil Improvement Association in 1912 could have imagined the changes that their organization would experience during the postwar years. As originally conceived, the Soil Improvement Association was concerned with adult agricultural education and demonstration work with an eye toward maintaining soil fertility and regional prosperity through the application of modern, scientific farming techniques. There was little indication during its formative period that it would move beyond it original mission and sponsor such a broad range of social activities, political programs, and business ventures that were important to farmers but that generally fell outside the arena of education as conceived by the early Soil Improvers. In fact, men like Eckhardt, who represented the original idealism of the farm advisor movement, were opposed to the sort of broader financial and social services the Farm Bureau was regularly offering the community by the end of the 1920s.

It is perhaps understandable, however, that far-sighted and prosperous farmers quickly realized that their powerful new organization offered great potential to improve their lives beyond educational opportunities. The momentum for farm advisor work originated, after all, in a desire to modernize American agriculture, and it became evident that improving the farm economy and raising living standards of rural Americans demanded services tailored to farmers' needs. The dual trials of war and agricultural depression created new responsibilities for the Soil Improvement Association and made it even clearer that education alone could not solve all the problems that confronted farmers. The Farm Bureau needed to undertake measures that would insulate farmers from the worst of the economic downturn. Men like Roberts represented a new generation of pragmatic farm advisors willing to experiment with new programs and policies in order help their constituents. While the Bureau continued to perform its educational functions that remained the core of its responsibilities, by tackling new problems that faced local farmers, and the nation, it evolved into a powerful collective voice for DeKalb County farmers in dealing with other interests and in confronting the challenges of farming in the modern era.

4 HARD TIMES

THE FARM BUREAU

DURING DEPRESSION AND WAR

•   The 1920s was a difficult period of transition for DeKalb County farmers. The postwar agricultural depression hit them hard, but as the decade wore on the economic climate stabilized. Although few farm incomes had returned to the heady wartime levels, the nation's industrial and business economies were profitable. Public rhetoric about the desperate conditions down on the farm lost its urgency. The natural flow of market forces was gradually readjusting the national agricultural sector to meet the new domestic and international market conditions. Land prices, which had inflated during and immediately after the war, were slowly falling to more reasonable levels. The economic downturn of the decade forced at least 160,000 marginal, overextended, or inexperienced farmers into bankruptcy and out of the farming business nationwide. An epidemic of rural bank failures, including a few in DeKalb County, had reduced the number of small, local, and marginal financial institutions, but the survivors were stronger and better capitalized. The availability of farm credit improved at both the local and national levels. Those farmers who weathered the economic shakeout borrowed money to absorb their neighbor's farms and restructured their own debts, either by renegotiating their short-term debts or, less frequently, by paying off principal. National farm mortgage debt increased about 10 percent, to 42 percent, between 1920 and 1930, a figure that suggests many farmers were borrowing extensively to hold on to their land and improve their operations.[1]

While stagnant economic conditions haunted American farmers, material goods that became widely available during the 1920s made daily life on the farm more pleasant than it had ever been. Rural Americans benefited from new communications and transportation systems. Telephone service, while still a luxury in private homes, connected neighbors and families. Radio, however, fundamentally altered the fabric of rural life. The relatively inexpensive device empowered farmers by providing practical information, such as weather reports, market news, and educational features. Radio also offered a world of entertainment, religious programming, and political messages that were not available to farmers just a few years earlier. "But what would we do without radio?" asked humorist Will Rogers.

> At any hour of the day or night, tune in and somebody is telling you how to live, how to vote, how to drink, how to think, when and how to see your doctor, and when to see your priest, and when to see your preacher, and how to put on fat, and how to take fat off, and how to make the skin stay white and how to make it stay black. Honest, no other nation in the world would stand for such advice as that. But we do and we like it. So the only thing that can make us give up radio is poverty. The old radio is the last thing moved out of the house when the sheriff comes in. . . . It is the best invention I think that has ever been.[2]

The automobile also revolutionized rural life during the decade. In 1920, about 30 percent of farm families owned cars, but by 1930 nearly 60 percent owned cars, despite the decade's difficult economic conditions. Road construction, sponsored by the federal, state, and local governments, as well as by private road improvement associations like the Lincoln Highway Association, enhanced marketing opportunities and made travel easier and more convenient for rural Americans. Many used the new roads to leave rural America far behind. The outflow of young people from the farms to the cities, about 6.2 million nationally during the decade, resulted in a decline of farm labor. Mechanization of farms offset that reduction in the farm labor force and even enabled farmers to increase their production while avoiding labor costs. Consequently, despite the sluggish economic conditions during the first half of the decade, the per capita income of farmers increased marginally during the second half. More important, the purchasing power of farm incomes was slowly rising, and, by the end of the decade, statistics show that it was near what it had been during the 1910–1914 period.[3]

World War I and its aftermath also changed farmers in less tangible ways. By the end of the 1920s, the industrialization of agriculture was largely complete on most of America's farms. Successful farmers had become businessmen, ruthlessly cutting costs and improving efficiency, dependent upon outside suppliers for much of the raw materials and tools of their living, accustomed to comfortable material lifestyles, and endlessly searching for customers who would pay cash for their output. It was a new agricultural reality that rewarded expertise in technology, science, finance, business, and education above practical farming experience and homely, rural wisdom. The postwar economic crisis taught farmers important lessons, especially the fact that prosperity in the modern era could not be taken for granted and was potentially just as fleeting as it had been for their grandfathers. The period saw increased political and social control in rural America. Bureaucracies of all kinds, from small local associations and governments to state and federal agencies, exercised influence over country communities. Farmers during the 1920s changed the ways they looked at their own lives. Increasingly, they compared their living standards, often unfavorably, with those of their urban contemporaries. Farmers also developed new outlooks about politics. It was clearer than ever before that decisions made beyond the township or county might drastically impact farmers' everyday lives. Isolation, parochialism, and individualism that had been hallmarks of the America countryside for generations eroded, and rural institutions, practices, and expectations reflected, to a greater extent than ever before, urban, modern trends of the dominant national society. City folk, too, changed their views of rural folk. Farming was increasingly a distant occupation, far removed from the bustle of urban lives. Where Progressive-era reformers once saw farmers as icons of American civilization worthy of preservation, during the 1920s city folk were as likely to view farmers as "buckwheats" and "rednecks"—quaint, sometimes distasteful, reactionary, backward-looking relics of an earlier time who stubbornly resisted the onrush of modernity.

Perhaps most important, hard times caused farmers to appreciate the power of collective action. The number of farm cooperative associations and farm bureau purchasing pools across the country increased, and they become more effective throughout the decade, reducing the costs of materials and exerting a stabilizing influence on prices. Collective action could also be used to sell farm products. For generations, farmers had been concerned primarily with production. The early farm advisors, like William Eckhardt, believed that the most effective way to improve the material lives of farmers was to boost farm efficiency and yields. They devoted themselves to improving farm output but thought less about how farmers' better yields could be sold. Farmers assumed, with some justification, that marketing was largely out of their control. With the modernization of agriculture during the opening decades of the twentieth century, however, farmers were producing more than ever before. In the new agricultural reality of the 1920s, farmers generally continued a production-side strategy of boosting efficiency, slashing costs, and maximizing harvests. Farm production remained strong during the decade, and even increased slightly, making more farm output available for market. Paradoxically, however, strong and efficient production did not immediately translate into better farm income because there was no corresponding increase in demand. Consequently, prices for most farm commodities, especially grain, were stagnant throughout mid-decade nationally and in DeKalb County.

Farmers could no longer afford to ignore the marketing phase of their operations if they hoped to translate their increased yields into greater incomes. As the agricultural economy struggled with the economic frustrations of the 1920s, the marketing issues took on an even greater importance. Just as the most successful farmers evolved into agricultural businessmen, innovative farm advisors and farm bureau leaders came to understand that their roles, by necessity, had to include matters of agricultural business as much as scientific farming techniques. Farm bureaus, which often included local businessmen and bankers as members, were at the forefront of these experiments in cooperative action during the 1920s. Many farm advisors, like Thomas Roberts in DeKalb County, considered marketing and businesses activities to be as important as their agricultural education programs. That focus on collective business activity would prove invaluable during the coming decade.

When Herbert Hoover was sworn in as president in March 1929, the nation's future looked promising. The "Great Engineer" had been elected by one of the largest pluralities in history, signaling that most voters approved of the business-oriented Republican policies that defined the political landscape of the decade. Commerce and industry flourished. Although economic sluggishness in the agricultural sector cast a cloud over that national economic success, farmers appeared to be emerging from the twilight of rural depression. Americans were optimistic and had every reason to believe that good times would continue during Hoover's administration.

During the spring and early summer of 1929, Congress debated one last attempt to craft agricultural relief legislation that would solve the farm belt's economic problems once for all. The new proposals looked to farmer-controlled cooperative associations as the best means of solving the farm problem. The Agricultural Marketing Act of 1929 stated that it was the policy of Congress "to promote the effective merchandising of agricultural commodities in interstate and foreign commerce, so that the industry of agriculture will be placed on a basis of equality with other industries, and to that end to protect, control, and stabilize the currents of interstate and foreign commerce in the marketing of agricultural commodities and their food products."[4] To achieve these ends, the president was empowered to appoint a board of eight members, approved by the Senate, to distribute $500 million in loans, with special emphasis on loaning money for cooperatives. The new law also provided for the creation of price stabilization corporations to control any surpluses that might arise, as well as for loans to those corporations to purchase those surpluses. The plan provided a balancing mechanism premised on the beliefs that the agricultural sector was on the road to recovery and the nonagricultural economy was essentially strong. The best the federal government could offer financially stressed rural America was to establish a mechanism for buffering the boom and bust cycles that might result from harvests that were either too good or too bad, seasonal variations in delivery times, international market conditions, and the like. The Agricultural Marketing Act was less than many farm organizations hoped to achieve after nearly a decade of legislative struggles in Washington, but they accepted it as the best they could get from unsympathetic legislators and hoped that it would help finally end the difficult economic problems of rural Americans.

The underlying assumptions that supported the bill, however, were flawed. The national economy during the summer of 1929 was less vigorous than it had been mid-decade. Business inventories in many sectors, for example, were increasing, which indicates that consumer purchasing power was not able to keep up with production. Important raw material industries were producing at reduced rates, and unemployment figures were creeping up as those companies laid off idle workers. Construction slowed, and automobile manufacturers were finding it harder to sell cars. Speculation had driven the prices of American stocks far out of proportion to the underlying worth and promise of the companies. The effervescent American stock market drew funds, including gold, out of unstable European economies still struggling with the financial debts of the war. Europeans invested in American stocks instead of durable goods and farm produce, exacerbating international balance of trade problems. Technological innovations reduced manufacturing costs and boosted business profits, but wages and prices did not keep pace with company earnings. Real wages stagnated, undercutting the vitality of the decade's consumer-driven economy. Personal debt was perilously high as working-class Americans borrowed to buy the products a modern lifestyle entailed. Throughout 1928 and early 1929, the Federal Reserve Board attempted to curb speculative borrowing by raising the discount rate and issuing securities they hoped would absorb some of the funds destined for stock investment. Against the glare of the nation's decade-long economic success in the business and industrial sectors, however, few noticed the symptoms of financial trouble, and fewer still would have believed they indicated anything more than temporary dislocations in a fundamentally sound economy.

The collapse of the stock market in the autumn of 1929, then, shocked Americans. While the stock crash by itself did not cause the Great Depression, it was the first and most spectacular manifestation of it and signaled the beginning of the most severe, dramatic, and prolonged economic crisis in the nation's history. The Great Depression was not a domestic phenomenon, but a worldwide economic disaster. The collapse of the stock market and the deflationary economic cycle erased equity and impaired principal, and tens of thousands of investors, who had seen their investments soar to fantastic levels just months earlier, watched in panic as their profits vanished and they plunged into bankruptcy. As the world's leading economy slid into depression, business, investor, and consumer confidence evaporated. The national climate of optimism was replaced, almost literally overnight, by one of caution and anxiety as everyone refocused their attention on how to salvage the most from the financial disaster. Demand fell for nearly all goods and services because buyers curtailed their spending, especially on nonessentials and expensive durable goods, such as cars and appliances. Retailers, sinking under the weight of unsold inventories, cut prices and cancelled factory orders. Industries slowed production, fired employees, stopped purchases of raw materials, and halted capital investment. Wages and prices fell while unemployment surged. As confidence in the overextended banking infrastructure declined, depositors withdrew their savings and stored them in safe deposit boxes, or under mattresses, further reducing the amount of capital available to restart the nation's financial engine. The country, and the world, descended into a deflationary spiral so severe and overwhelming that financial and business leaders were powerless to stop it. Elected officials were even more helpless in their futile attempts to articulate a comprehensive, or even rational, political response to the disaster.

The impact of this new financial crisis, coming so close on the heels of the difficult economic conditions of the 1920s, was disastrous for farmers. In earlier periods of the nation's history, when the majority of Americans were farmers, agricultural depressions often triggered depressions in the commercial and manufacturing sectors. In 1929, that pattern was reversed, and what began as an industrial depression soon engulfed the countryside. Demand for farm products declined as Americans engaged in ruthless underconsumption to make ends meet. Americans ate less and changed their eating habits, substituting cheaper foods for more expensive ones. In 1932, Americans spent roughly 40 percent less on food than they did three years earlier. Clothing expenditures declined similarly as consumers bought fewer garments and repaired and recycled those that they already owned. Foreign markets vanished when importing nations fell into economic hard times. Tariff barriers sprang up at home and abroad as nations struggled to support their flagging producers during the crisis. Deflation was especially hard on debtors. Farmers who had borrowed heavily during flush times and restructured their loans in response to the economic problems of the 1920s found that those loans could not so easily be paid off when farm and commodity prices were falling. Banks and insurance companies that held farm mortgages were anxious to improve their own financial solvency and were not inclined to renegotiate, extend, or make new loans to farmers who were already heavily indebted. Moreover, foreclosure was hardly an appealing incentive for lenders to make new loans during a time when collateral values were rapidly declining. In major industries, like steel and automobiles, producers had control over labor and supply and were able to exert some pressure in the market to maintain prices, albeit at drastically reduced levels. Farmers, however, had little control over their fate. There were simply too many small farmers to effectively restrict output and boost prices. Business could easily halt production and fire employees, but farmers had significant fixed costs, operated on a time schedule dictated by growing seasons, and were at the mercy of a host of natural and man-made measures that were beyond their control. In 1931, for example, ideal growing conditions produced bumper crops of cotton and wheat that flooded the markets, driving prices for those commodities down even farther—corn tumbled from about $0.77 per bushel in 1929 to $0.32 three years later, and wheat from $1.05 to $0.38.

Between 1929 and 1932, farm income fell 60 percent nationally. In 1929, farm incomes had recovered from the decade's agricultural recession to about 92 percent of parity, but were only at 58 percent of parity in 1932. At the start of the Great Depression, farm incomes amounted to only about 70 percent of manufacturing wages. Three years later, farmer incomes had fallen to only about 50 percent of average manufacturing earnings, which themselves had already taken a drastic reduction. The total value of agricultural capital value fell from about $58 billion in 1929 to $36 billion in 1933.[5] Statistics, however, cannot fully convey the human dimensions of the crisis. In many parts of rural America, farmers were desperate. Families cut all but the most necessary expenditures. Many returned to a self-sufficient lifestyle, more reflective of earlier farm generations, that included canning, keeping dairy cattle, home slaughtering and meat preservation, poultry raising, and cottage industry. Farm women, especially, bore the burden of these adjustments. Farmers fired their laborers, shifting more work onto family members or doing without help. Some reverted to mules and horses because they were cheaper than tractors; many shared equipment with neighbors. The inability of farmers to raise cash to pay their taxes resulted in foreclosures and tax sales. It also meant that community services declined. Public infrastructure, such as roads and bridges, fell into disrepair. Local schools cut back on teacher salaries and other expenses, or closed altogether.[6] The shortage of cash in the countryside also severely impacted local merchants and service providers. Barter and other informal economic practices, which had always been a fairly routine country practice, became even more common for farmers and townsfolk alike. The small Waterman, Illinois, hospital, for example, sometimes accepted corn as payment for medical services.[7]

In the summer of 1929, however, such dire conditions lay in the future, and DeKalb County farmers were slowly adjusting to the agricultural economics of the 1920s. During the decade, DeKalb County fared better than many counties in the corn belt and lost only about a hundred farms, bringing the total number to around 2,300. The Farm Bureau continued its programs and services to area residents. The insurance business was booming, and drivers found, just as predicted, that the money they saved in their auto insurance rates often paid the price of their Bureau membership. The farm casualty and life insurance programs were equally successful. The collective Dairy Marketing Association in Waterman was cooling between 20,000 and 30,000 pounds of milk daily for the Chicago market and paid dairymen about $2.45 per hundred pounds of milk, an amount considerably higher than the $2.00 they earned two years earlier. Improved roads meant that trucking firms had the ability to reach the countryside, and the Bureau was educating area stockmen and regional trucking companies and independent drivers about the virtues of cooperative marketing of livestock. The oil-marketing company was very popular and, in 1928, paid Farm Bureau members nearly $3,000 in total rebates. Farmers still asked for soil testing, and Tom Roberts persisted in experimenting with new varieties of crops. The Bureau's financial services were popular, and increasing numbers of farmers relied on Roberts and others in the Bureau to help them negotiate loans, fill out tax forms and board of health documents, and train them about farm record keeping and management. The DeKalb County National Farm Loan Association, founded in 1919 by the Soil Improvement Association, was busy refinancing loans. Hundreds of children annually participated in Bureau-sponsored 4-H programs, and many of DeKalb boys and girls enjoyed great success in local and state level competitions. In November 1928, the Bureau ran a corn show that awarded a $50 silver cup to the winner, $25 in cash prizes, a $15 hog feeder, a ton of coal, and even $15 in new underwear—the eighteenth-place prize was a jug of cider.[8]

Underlying all this success, however, was the very real problem that DeKalb county farmers simply were less prosperous than they had been during the war years. Farm Bureau membership rebounded from its low in 1922 to 1,250 members in 1928, but that was still far short of the peak membership of more than 2,000 it enjoyed during the immediate postwar period. The Farm Bureau planned an aggressive membership campaign for 1930 and 1931, but the economic uncertainty of the time meant that fewer and fewer farmers found themselves willing to pay the $10.00 local dues.[9]

Perhaps most important for the Farm Bureau, however, was the fundamental change that occurred in its relationship with governmental agencies. For nearly two decades, beginning with the DeKalb County Agricultural Association seed business, the Farm Bureau had been steadily increasing its commercial activities. Rank-and-file members considered the Bureau's various affiliated enterprises, such as the DeKalb County Agricultural Association, the oil company, and the insurance brokerage company, as integrated parts of their Farm Bureau designed to address specific local needs, promote the best interests of area farmers, and ensure prosperity for the region. Moreover, some of the operations even returned dividends to members. Roberts helped organize most of them and had a hand in managing nearly all, and nobody in DeKalb County even considered that there might be conflicts of interest between the Bureau and its related businesses. Managing such commercial activities was simply inherent in a farm advisor's duties.

State and federal agencies, however, took a different view. While some of the operations, like the petroleum delivery and insurance affiliates, could be interpreted as cooperative services of the kind that farm bureaus provided, as early as 1927 government bureaucrats and legal advisors were concerned about the close relationship Roberts had with the DeKalb County Agricultural Association, which was, after all, a publicly-held corporation. The Farm Bureau originally hired Roberts pursuant to a memorandum of understanding with the University of Illinois, as representative of the extension service, to provide educational services as the Farm Bureau agricultural advisor. The Farm Bureau, according to the extension service, was simply "an educational and advisory agency in the most practical sense of the term."[10] As the decade wore on, however, Roberts spent increasing amounts of his time working at the DeKalb County Agricultural Association. That, according to the extension service, was a violation of the agreement and led to a rift between the Farm Bureau and the extension service. In mid-summer 1929, federal and state agricultural agencies warned that they would end payments to the DeKalb County Farm Bureau for agricultural extension work, citing the disproportionate amount of time the farm advisor spent on the commercial projects instead of the educational programming agreed to in the memorandum. It would be a serious financial blow to the Farm Bureau. There was also the fear that such a move might undermine the Bureau's legitimacy and jeopardize its affiliated commercial programs. Perhaps more important, Bureau leaders confronted, once again, the difficult issue of their organization's independence. For the first time since the earliest days of the Soil Improvement Association, the Farm Bureau faced a future totally reliant on its members to fund its programs.[11]

In the midst of the controversy, the Farm Bureau hired Russel N. Rasmusen as assistant farm advisor in August 1929. In reality, he was hired to become the acting farm advisor for Roberts, who was by then spending nearly all his time at the Agricultural Association. When it became clear that the government agencies were not willing to reconsider their financial support as long as Roberts was primarily engaged with the Agricultural Association, he stepped aside in April 1930. "[T]he University Extension people thought I was too commercial," said Roberts, "fooling around with seeds, oil, Land Bank loans, and the service company, so they withheld their money and we went ahead and became quite successful anyway."[12] The Farm Bureau board of directors appointed Rasmusen as interim advisor, and shortly afterward they made him their new county advisor. Ironically, although original leaders of the Soil Improvement Association believed that their new organization could operate best if it was free from the political tethers that might accompany public funding, by the end of the 1920s the Bureau had become so dependent on government financial assistance that Rasmusen's first order of business was to work with the state and federal agencies to restore the funding the Farm Bureau desperately needed for its core educational and service activities. His timing was poor. Within a few months, the nation plunged into the Great Depression. Suddenly, the strategies that the Farm Bureau had inaugurated during the 1920s to address the agricultural recession of the decade, the goodwill of area farmers, and the commitment of Bureau leaders and members to their community, would become all that much more important in dealing with the new economic crisis.[13]

The most pressing problem for area farmers at the beginning of the Great Depression was securing credit at a time when markets vanished, banks teetered toward insolvency, equity evaporated, commodity prices plunged, and property and machinery values dropped. During 1930, the first full year of the Great Depression, hundreds of DeKalb County farmers needed to refinance their loans to save their farms from foreclosure. Between 200 and 300 distressed farmers looked to the Farm Bureau–affiliated DeKalb County National Farm Loan Association for assistance. The demand was so great that the Association eventually loaned more than $6 million on some 500 farms. Local banks also relied on the Association to help contract mortgages and stabilize the credit crisis. Banks in the county, however, were also facing difficult times. As once secure county banks seemed to be faltering, area residents withdrew their savings— bank deposits in the county declined between 40 and 50 percent. Rumors about the insolvency of area banks resulted in bank runs that further undercut the institutions. Three area banks—the Rollo Bank, the First Trust and Savings Bank of DeKalb, and the First National Bank of Sycamore—failed under the stress. The Farm Bureau itself lost some of its deposits when the Sycamore bank collapsed, a loss that made weathering the difficulties of an uncertain future even more difficult. Roberts called an emergency meeting of DeKalb bankers in late January 1930 at the Farm Bureau office to develop a strategy to address the developing banking crisis. The central problem, they concluded, was that depositors were withdrawing deposits at the very time the banks needed cash to lend and remain in business. To stop the local bank runs, restore depositor confidence in their institutions, and increase their solvency, the bankers declared a bank moratorium and closed their doors to business. That action ended the wild rumors of impending bank collapse and frenzied pace of withdrawals. As a consequence of their prompt action, bank deposits stabilized and investors slowly returned their deposits, making more money available for banks to loan to area farmers and businesses to keep them operating. More important, such cooperative action helped save the remaining cash-strapped DeKalb County banks, which, in turn, provided area farmers and businesses with the credit they needed to weather the Depression.

Although the Great Depression seriously eroded the economic well-being of DeKalb County farmers and forced some off the land, as a whole they were in a much better position than many other farmers in America to endure the downturn. In the spring of 1931, DeKalb ranked sixth in total crop and stock values of all Illinois counties. Crop yields were consistent. Houses and barns remained painted, farm equipment and cars were maintained, electricity lit most farmhouses, telephones stayed connected, schools taught children, radio antennas brought broadcasts to even the most isolated homes. Land values slipped, but remained relatively stable throughout the decade. Land in the county was selling at around $150 per acre, certainly down from the heady World War I prices of $350 per acre, but better than many other places in the nation—the Sycamore Preserve Works, a local cannery, was paying between $5 and $10 per acre in rent to raise its produce, prices that could purchase land in vast stretches of the Central Plains states. Equally important to the survival of DeKalb county farmers was the fact that they were less likely to lose confidence in themselves. "Things on DeKalb community farms are not perfect, there is no use saying they are," wrote one local editor, "but they are far from hopeless and the best feature of it all is that these farmers about here are working out their own salvation and [not] standing around on street corners yelling for someone to come help them. With that kind of spirit DeKalb farmers are going to regain their deserved place."[14]

Nevertheless, the Great Depression placed severe strains on the Farm Bureau. Membership remained a serious concern. It fell to just under 1,000 members by 1935, as more and more county farmers found it impossible to afford the $10 annual dues. In 1932, the financially distressed DeKalb County Board of Supervisors stopped paying the $4,000 to support advisor services that it had allocated since the earliest days of the Soil Improvement Association. In the autumn of 1932, the Farm Bureau began to cut services, including reducing funding for speakers and meetings, club work, office expenses, and Bureau publications. It fired its assistant farm advisor, W. Howard Kauffman, just a year and a half after he started work, and reduced the salaries of the Bureau's two secretaries. Even these drastic measures, however, were not enough to keep the Farm Bureau in the black. The Bureau's net worth by 1935 was only about $15,000, and it was losing nearly $600 a year. Yet the DeKalb County Farm Bureau was better off financially than its counterparts in many Illinois counties, and during the decade it remained one of the strongest bureaus in maintaining membership and programming.

It continued to be successful because, although the Farm Bureau cut items from its budget, it did not abandon its core mission of agricultural education while it struggled to keep operating. Soil testing had always been an important Farm Bureau function, and Rasmusen instructed farmers in sophisticated new techniques for maintaining soil fertility. Farmers learned to apply limestone, for example, in varying amounts to different parts of individual fields, instead of treating an entire field uniformly as had been the more common practice earlier. Erosion on sloping fields was becoming a problem, so in 1931 the Farm Bureau sponsored a demonstration to increase awareness and interest in erosion and how to control it. Beginning in 1932, Rasmusen and the Bureau spearheaded a campaign to eradicate thistle and quack grass, both noxious weeds that interfered with the county's commercial crops, through the use of smothering crops, like millet and rye, as well as mechanical and chemical techniques. The Farm Bureau, in cooperation with the county weed committee and the Illinois Department of Agriculture, continued weed eradication efforts in the following years. Rasmusen even used the radio to broadcast information on weed control to area farmers. When chinch bugs invaded DeKalb County following the drought of 1934 and threatened to destroy the county's crops, hundreds of area farmers attended Farm Bureau demonstrations about controlling the pest. The Bureau also distributed four rail tankers of creosote to 700 county farmers to use as chemical barriers to help prevent further spread of the infestation.

The Bureau also helped farmers further improve their crop production through the introduction of new varieties of improved seed. The most important advancement was the introduction of hybrid corn. Until the mid-1930s, farmers relied on open-pollinated corn seed, carefully selected and graded for maximum yields in the northern Illinois soils and climate. In 1921, Roberts became interested in the technique of selectively cross-breeding and inbreeding corn, a process that produced hybrid seed, as a method of boosting corn yields. Beginning in the spring of 1925, the DeKalb County Agricultural Association's scientists, including Roberts and Charlie Gunn, and later Ray Nelson and Orton Bell, quietly experimented with hybrid corn. During their early trials, Roberts and Gunn did not even inform the Agricultural Association's board of directors about their work, in part because they believed that the work of placing paper bags over corn tassels, a step required to prevent unwanted airborne pollination, seemed too juvenile for grown, professional men. Four years later the work paid off, and Roberts announced that they had succeeded in breeding corn that would outproduce the best open-pollinated seed corn varieties.[15]

Making the transition from the laboratory to the marketplace, however, took some time. As late as 1933, the Association's hybrid project had still not reached the stage where hybrid seed was available in large quantities at a cost that was competitive with the best open-pollinated varieties. In 1934, drought damaged nearly all the precious hybrid seed plants and interfered with the plant's carefully managed pollination cycle, and the Association salvaged only about 325 bushels of seed corn, most of which went to a handful of DeKalb County farmers. Mastering the technique of raising hybrid seed, however, was not the only problem. The new hybrid seed was clearly superior and promised better yields than open-pollinated varieties, but when the hybrid variety finally arrived in marketable quantities in 1935, area farmers were slow to adopt it.[16] It was one thing to produce fine results on experimental plots, but skeptical farmers demanded practical evidence that the new seed was worth its extra cost, especially during the Depression when every dollar meant so much to farm families. One area farmer was so cautious about hybrid corn that he planted only a small plot in the middle of his corn field and surrounded it with another variety—just in case the hybrid failed, he wanted to be certain that none of his neighbors found out he was dabbling in the new seed![17] Hybrid corn seldom failed to outproduce other varieties, however, and as the new hybrids became more widely available and proved their worth, more and more farmers adopted them. The Agricultural Association's seed representatives also aggressively marketed the new product, and the Farm Bureau regularly informed its members about the benefits of hybrids. In 1935, the first successful year for the Association's hybrid seed program, it produced and sold 14,500 bushels of seed. By the late 1930s, hybrid corn, while it did not completely replace open-pollinated seed in DeKalb County, was playing a significant role in the county's agriculture output. For the

DeKalb County Agricultural Association, the success of its hybrid seed kept the company afloat during the Depression years and laid the foundation for one of the most important domestic agriculture companies of the twentieth century. The close relationship between the Association and the Farm Bureau also meant that the success of the corporation would have a profound influence on the Farm Bureau for the next five decades.[18]

Although soil and crop education were traditionally the most important Farm Bureau goals, as early as the 1920s the Bureau was equally supportive of the county's livestock producers. "I preached to the farmers," said Roberts, "that hogs paid for more farms in DeKalb County than anything else."[19] During the Depression, the Bureau remained committed to the stockmen. It sponsored University of Illinois specialists to teach courses on new livestock management and feeding techniques. Dairy programs introduced the most economical production methods. Livestock farmers especially welcomed the attention, and, during the Depression, though Bureau membership was low, membership in the Farm Bureau's Dairy Herd Improvement Association increased. The Farm Bureau set up a cattle-marketing committee to help educate stockmen about new sales techniques and introduce them to regional beef buyers, especially in the Chicago area. Sheepmen benefited from the Bureau's support of their collective DeKalb County Wool Pool. Poultry programs stressed the practical importance of winter feeding and management, culling, and business accounting. The Farm Bureau used its collective purchasing power to become the county's leading supplier of hog cholera serum and virus in 1931. The Bureau also provided veterinarians to teach hog farmers how to vaccinate their own herds. At a time when some farmers were returning to horses as a source of farm power, the Farm Bureau, in 1932, sponsored a demonstration program on how to use large teams, with as many as eight horses, to pull farm implements—a skill that had once been common but that many farmers had forgotten, or never learned, with the widespread adoption of tractors. To protect the animals, Rasmusen called on county veterinarians to ask for their cooperation in systemic control of botflies that were threatening to take a heavy toll on county horses. He eventually arranged for half of the horses in the county to be tested and treated, and the project eliminated the botfly infestation by the end of 1933. To help farmers reduce their living expenses, the Bureau even arranged demonstrations to teach farmers how to butcher and preserve their own meat.[20]

Crop and livestock programs formed the core of the Bureau's programming, but the Bureau found that the Great Depression spurred members' interest in its social activities, especially youth programs. Support of 4-H clubs was a natural extension of its educational mission and provided rural youth the opportunity to succeed in agricultural and related fields while promoting leadership, citizenship, and cooperation. Youth programs provided important recreation for farm children and had the added attraction that they were inexpensive to run. The 4-H clubs were popular with both the children and their parents, in large part because farm families found that club programs offered rural children education in agriculture and home economics they could not otherwise afford. Perhaps most important, however, social programming enriched farm life and helped keep young rural men and women on the farm at a time when increasing numbers of them were leaving the countryside for jobs and lives in towns and cities.

Rasmusen especially recognized the value of social programs and encouraged club membership and recruited parents as local club leaders. By 1932, county 4-H activities had grown to more than 400 members participating in twenty-seven separate clubs. Children had a variety of projects from which to choose, but the most popular included livestock for the boys and gardening, food preparation, and clothing construction for girls. Beginning in 1933, the Bureau secured the calves for the calf projects and provided calf insurance for the exhibitors. In 1934, Rasmusen organized the DeKalb County 4-H Club Federation to better organize the clubs throughout the county. The local leadership was particularly strong, and several county children enjoyed local and regional success in the club. Three—Dana Schweitzer, Ralph McKenzie, and Donald Mosher—attended the national 4-H camp in Washington, D.C., during the decade, an honor awarded to only two children in each state annually. In 1936, Rasmusen and the Farm Bureau created the DeKalb County Rural Youth, a program designed to offer young adults who had graduated from 4-H club an organization where they could continue educational and community service projects. Rural Youth also provided important recreational opportunities, such as dances and picnics, where young rural men and women could meet and socialize under the supervision of chaperons—many Rural Youth alumni fondly referred to the organization as the "matrimonial bureau" because of the scores of county marriages between "RY" members.[21]

In addition to its 4-H programming, the Farm Bureau supported other social activities. It sponsored, for example, a "recreational school" where representatives of local organizations learned to organize community entertainment projects. It was instrumental in organizing a county competition of one-act plays and musical quartets. The Bureau encouraged athletic competitions in the county to provide area youth the opportunity to compete in a variety of sports. One of the most exciting athletic programs was the Farm Bureau amateur baseball team, which was organized in the spring of 1931 to compete with other Farm Bureau teams from counties in northern Illinois. In 1935, the Bureau helped organize and run the county's first corn-husking contest and, a year later, hosted the state husking contest. The Bureau also sponsored women's, and later men's, softball teams in conjunction with the Illinois Agricultural Association's annual Illinois Farm Sports Festival, held in Champaign. A year later, in 1936, the Bureau hosted its first annual Farm Sports Festival, an event that encouraged competition in both team and individual sports. Bureau-sponsored sports programs were very popular and successful. The baseball rivalry between DeKalb County and neighboring Ogle County was especially fierce, and the level of play was remarkably high—the DeKalb Farm Bureau team won third in the Illinois State Farm Bureau League competition in 1937 and won the Illinois Farm Bureau state championship and interstate championship with Iowa in 1939 and 1940. George Tindall, the star left fielder and shortstop of the team, briefly played semiprofessional baseball before returning to his family's farm. DeKalb county won the high-point sweepstakes award at the Illinois Agricultural Association's Farm Sports Festival in 1938. The DeKalb women's softball team won the state championship four consecutive years between 1937 and 1940 and went on to defeat the Iowa state women's champion team in 1940. The over-thirty-five softball team won the state title in 1938 and 1939 and also defeated the Iowa champions in 1939.[22]

Educational and recreational activities gave youth opportunities to socialize and improved the quality of life in DeKalb County, but the anchor of farm life remained the farm home. Traditionally, the farm wife was responsible for the smooth, efficient, and economical operation of her household. During the uncertain times of the 1920s, and again during the Depression, farm wives found that job to be even more demanding. Although it was not a new or unique idea to offer homemaker extension services, and many Illinois counties already benefited from home extension by the middle of the decade, the DeKalb Farm Bureau historically showed little interest in providing it. In the early days of the Soil Improvement Association, in fact, William Eckhardt and the Association leadership were adamantly against supporting social activities, especially a home advisor program. During the economic troubles of the 1920s, however, such ideological opposition to a home advisor began to soften.

In April of 1922, a representative from the home economics department at the University of Illinois met in Waterman with DeKalb County residents and outlined a county organization plan for a home advisor. She explained that the Home Bureau was similar to the Farm Bureau for men and that the home advisor worked for farm women and families just as the farm advisor assisted farmers. She also noted that eighteen Illinois counties already had permanent home advisor programs, three of them dating back to 1919, and six counties employed a home advisor on a temporary basis. Moreover, the Smith-Lever Act allocated $1,200 a year to counties to help pay for home advisors. Educational topics for home bureau programs already in place in other counties included family nutrition, home management, health and sanitation, interior decorating, and rural school hot lunch. The DeKalb residents who attended the meeting voted unanimously to form a home bureau organization. The DeKalb County Agricultural Association agreed to lend its support to convince the Soil Improvement Association directors to adopt a home advisor plan. Although Tom Roberts was far more receptive than his predecessor to the idea of a home advisor and lobbied the Farm Bureau board of directors for one, the directors did not agree and the county remained without a home advisor until 1936, when Rasmusen convinced the board that the farm home was critical to the success of the farm. The well-being of county farms, and the Farm Bureau itself, depended on farmers' quality of life as much as on their economic success. Despite the financial hardships of the decade, the Bureau directors finally agreed to support home extension in DeKalb County.[23]

Rasmusen and other supporters of the idea asked local women to join the new DeKalb County Home Bureau (later called the Homemakers Extension Association). Interest in the new program was high, and 378 women signed charter membership cards for the DeKalb County Home Bureau during its initial membership drive. By the time of the Farm Bureau annual meeting in February 1937, plans were well under way for the new Home Bureau. The Home Bureau members held their organizational meeting on March 24, 1937, at the DeKalb Methodist

Church, where they elected officers and formally accepted the state nonprofit corporate charter, constitution, bylaws, and budget for the new organization. It began operating out of its offices in the Farm Bureau building. In May 1937, DeKalb County welcomed its first official home advisor, Charlotte Biester. Biester was originally from neighboring Belvidere, Illinois, but had extensive home economics education experience in North Dakota, South Dakota, Kansas, and in the Indian Field Service for the U.S. Department of Interior. Although DeKalb County lagged behind other Illinois counties in creating a Home Bureau—it was the fifty-first Illinois county to do so—by the end of the first year, DeKalb County ranked tenth in the state in Home Bureau membership.[24]

The fundamental goals of Home Bureau members was to make their homes "economically sound, mechanically convenient, morally wholesome, artistically satisfying, socially responsible, spiritually inspiring, founded upon mutual affection and respect." "Home," members proclaimed, "should be the center of every woman's interest but not the circumference." Such sentiments reflected perfectly the Farm Bureau's mission of enriching the lives of rural Americans. Yet it was also a new direction for the DeKalb County, Farm Bureau, and an especially big step for the women of the county who were far more at ease in church sisterhoods, school parent organizations, local interest clubs, and community issue committees. "The future of the DeKalb County Farm Bureau lies in the hands of its members," wrote Mrs. Willard Cook, Home Bureau advisor for McHenry County, to the 1937 commencement ceremony of the first DeKalb County Home Bureau training school.

> When we have a task to do which seems to us to be especially difficult, it is made easier by keeping the goal in mind. First of all, we must remember that this is our opportunity to get an adult education. We must expect to work at it just as we did our lessons in school. . . . Let us each do our part to the best of our ability, even though at first, our knees may knock. We just can't help others without helping ourselves. We'll find that as we continue to do our part, the task becomes easier, and the knocks will become more infrequent. . . . As you go to Farm and Home Week and hear the yearly reports of the various County Presidents, you will notice a great difference in their achievements. This is not only a record of their accomplishments, but also, indirectly, a report of the effort and sincerity of their membership.

Ward McAllister, of southern DeKalb County, and the results of an early test of DeKalb hybrid corn, October 31, 1933. The hybrid showed a 35 percent increase in yield compared with yield from open-pollinated Western Plowman corn.

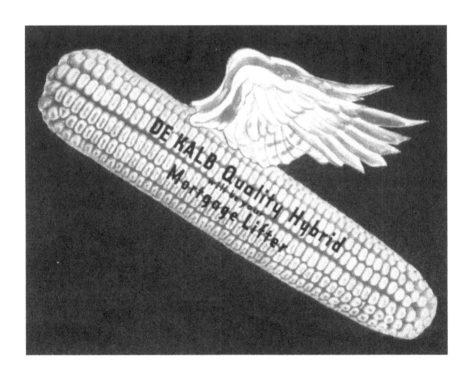

above—An early advertising image for the DeKalb Agricultural Association hybrid corn seed, mid-1930s. The winged ear represented the Association's advertising slogan that its hybrid corn was a "mortgage lifter."

right—Advertisement for the fifth annual Illinois Farm Sports Festival, 1940. The festival was sometimes referred to as the "barnyard Olympics."

DeKalb County Farm Bureau women's softball team, September 1937. This team won the 1937 Illinois Farm Bureau state championship. Front row (left to right): Edna Story, Lucille Larson, Lorraine Colby, Elizabeth Kientz, and Rebecca Colby. Middle row (left to right): Edna Schweitzer, Marion Ruth Challand, Gladys George, Marcella Williams, and Sally Spengler. Back row (left to right): Evelyn Mosher (captain), Hy Johnson (manager), Donna Shellaberger, Eugenie Donnelly, and Dorothy Bartlett. In the final round of play after a long summer of action, the women defeated Monroe County, the 1936 state champions, and Peoria County for the state title.

Some of the original founders of the Soil Improvement Association at the annual DeKalb County Farm Bureau–Home Bureau–4H Picnic and Sports Festival, held at the Sycamore Community Park, celebrating the twenty-fifth anniversary of the DeKalb County Farm Bureau, August 28, 1937. Left to right: John Blair, C. D. Schoonmaker, George Hyde, William Eckhardt, George Gurler, Aaron Plapp, Frederick B. Townsend, Henry H. Parke, Edger E. Hipple, George A. Fox, William F. Leifheit, and Orton Bell. This group of twelve men included seven farmers, two who owned farms, a newspaper editor (Schoonmaker, editor of the *Genoa Republican*), a dairyman (Gurler), and a farm advisor (Eckhardt).

The benefit of Farm Bureau membership, 1939. Throughout the life of the DeKalb County Farm Bureau, recruiting was an important way to sustain membership and raise funds for programming. That was especially true during the Great Depression, when Bureau membership plunged because of difficult economic conditions.

CHOLERA MUST GO!

American soldiers and workers need your pork. Make sure they'll get it. Vaccinate your pigs young with **FARM BUREAU SERUM** and **VIRUS.**

SEE YOUR FARM BUREAU

left—Hitler conspiring with the hog cholera devil to undermine the war effort, 1942. Hog cholera, a perennial scourge of swine farmers, was frequently represented in prewar Bureau advertising as a devil or the Grim Reaper stealing profits from careless farmers. In this case, the devil wants even more, and stopping cholera is elevated to a patriotic duty.

Flies are deadly "filth columnists" that can sabotage our increased food production. With every farm crowded with more cattle, hogs, and poultry, we have an ideal situation for new epidemics unless sanitation control is carefully exercised. This situation demands better control of flies. Give your livestock "anti-aircraft protection" against flies and insects with Blue Seal Fly Spray.

ASK YOUR BLUE SEAL TRUCKSALESMAN FOR

BLUE SEAL FLY SPRAY

PROTECTS CATTLE-KILLS HOUSE FLIES

Kishwaukee Service Company
DeKalb, Illinois

"Filth columnist" flies demand the "anti-aircraft protection" available from Blue Seal insecticide, 1942. The flies' menacing faces show typical wartime characterizations of the Japanese.

right—Bureau advertisement encouraging members to save rationed fuel, 1943.

below—The Farmer's Program for 1943 included utilizing Farm Bureau services.

Agricultural Achievement Award

TO THE FARM FAMILIES OF

DeKalb County

IN GRATEFUL ACKNOWLEDGMENT of services rendered to their country in its time of need, the Agricultural Achievement Award is hereby presented by the War Food Administration of the United States of America. Overcoming great difficulties by decisive action, laboring with determined devotion, joining together in making wise use of all their resources, the farm people of this county have answered their country's call for utmost food production. They have thus contributed in fullest measure to the cause of the United States of America and to the preservation of human freedom.

Given for the 1943 season,

Marvin Jones
War Food Administrator

Facsimile of the United States War Food Administration "A Award" awarded to DeKalb County in 1943. DeKalb County was one of only three Illinois counties to receive the honor. Henry Parke accepted the award on behalf of all DeKalb farmers.

above—Howard Mullins, a
Shabbona Township farmer,
was the president of the Bureau from February 1956 until
February 1974. He led the
DeKalb County Farm Bureau
to national prominence and
was the most influential
county farm bureau president
in history.

right—Elroy E. "Al" Golden,
the ninth farm advisor, 1956.
Golden was the advisor from
May 1954 until February 1970,
a period during which the
DeKalb County Farm Bureau
rose to national prominence.

Farm Bureau float in the
DeKalb centennial celebration,
1956. Henry Parke is honored
as the "Farm Bureau Originator," and the Bureau president,
Howard Mullins, stands at the
rear of the platform, behind a
photograph of the Bureau's
new building. Less than a year
later, on May 25, 1957, Parke
died at the age of eighty-one.

Thomas Roberts, Sr., the second farm advisor for the DeKalb County Farm Bureau, and Roger Rasmusen, admiring the hybrid corn Roberts helped pioneer in DeKalb County.

Soil testing was an important service for the Bureau since its beginning in 1912. Here, soil-testing technician Willard Cook is testing samples in the Bureau's soil laboratory, 1956. That year, Cook tested samples from 35,755 acres. The Bureau's modern equipment rendered results that were more precise than was possible using older methods, an improvement that saved DeKalb County farmers thousands of dollars in unnecessary soil additives. A typical soil prescription at the time was 3 tons of crushed limestone, 200–300 pounds of potash, and 1,500 pounds of rock phosphate per acre.

Biester herself was deeply involved in programming from the beginning and determined that the new Home Bureau succeed. During her first year in the position, she prepared 81 circulars, wrote more than 900 personal letters, made 555 telephone calls, prepared 262 newspaper articles, helped prepare 18 exhibits, and helped organize a style show, a county chorus, and a music and drama tournament. In addition to her regular appearances at project meetings held by the nineteen township Home Bureau units across the county, she attended the state sports festival, state 4-H camp, state farm and home advisors conference, and three district advisors conferences, and participated in thirty-three other meetings. She also took six and a half days of vacation. It was the programming, however, that made the Home Bureau such an instant success. Like the educational programming directed toward farmers, Home Bureau lesson plans stressed practical home lessons. During the first year, for example, members learned about a broad range of subjects including home furnishings, pressure cooker use and safety, the use of colors in the home, window treatments, nutrition, rural schools, home safety and first aid, "attractive accessories," vacuum cleaners, and Christmas suggestions. "Never have I attended a meeting," wrote Cook, "but what I felt I had gained from it and had become a better homemaker, because of it."[25]

While the Farm Bureau focused the majority of its attention on local programming, it also emerged as the leading voice for area farmers in national farm policy debates. Throughout the 1920s, DeKalb County Farm Bureau leaders supported the McNary-Haugen proposals, which were passed by Congress in 1927 and again in 1928 but vetoed by President Calvin Coolidge on both occasions. The new Roosevelt administration, however, took a different approach to solving the farm crisis. The result was the Agricultural Adjustment Act (AAA) of 1933, designed to raise the prices of farm crops and restore the farmers' purchasing power to pre–World War I levels. Unlike the McNary-Haugen and the Federal Farm Board's marketing approach to boost farm incomes, which probably would have been inadequate to solve the agricultural crisis of the Depression in any event, the guiding principle behind the AAA was the voluntary restriction of farm production and the payment of a subsidy to farmers who withdrew croplands from production. The AAA empowered the secretary of agriculture to levy a tax on processors to pay for the subsidies. By 1934, more than 40 million acres of farmland had been withdrawn nationally from production, for which farmers received hundreds of millions of dollars in benefit payments. The AAA also authorized volunteer organizations at the county level to administer the program on a local basis. In DeKalb County, the Farm Bureau played an instrumental role in educating farmers about the new law, and Rasmusen devoted considerable time to the program.

The first phase of the AAA program in DeKalb County began in the fall of 1933 and was directed toward reducing wheat acreage, a program that especially impacted the northern part of the county where the thinner timber soils were ideal for growing wheat. Rasmusen reviewed the Farm Bureau records and conducted a mail campaign that ultimately identified more than 400 farmers who grew wheat, at least occasionally, in the county. The local "County Campaign Committee," comprised of nine men, seven of whom represented local farm organizations, called a series of meetings in the county to present the AAA plan and recruit participants. Fifty-one farmers signed on immediately, and by the end of the campaign, 119 farmers, representing nearly half of the total wheat acreage in the county, had signed onto the acreage reduction agreements and received AAA benefit payments. A similar plan for the county's sugar beet farmers succeeded in nearly 80 percent of the beet growers signing federal agreements to curtail production.

The more important AAA program in the county, however, was government's corn-hog program. In the winter of 1933–1934, Rasmusen appointed nine representatives of area farm organizations to assist him with the educational meetings for the federal corn-hog program. In January 1934, Rasmusen and his new AAA assistant, Earl Wenzel, conducted more than thirty informational meetings with more than 3,000 local farmers to teach them about the new federal crop reduction benefits, answer questions, and convince them to sign up for the program. By the late spring, 1,842 farmers had agreed to participate in the federal program, representing about 97 percent of the hog producers in the county. Rasmusen, in the fall of 1934, polled the hog farmers again at a referendum meeting to ascertain the success of the federal program, and nearly three-quarters of them voted that they were interested in continuing the program in 1935. In addition to the assistance it provided with the AAA, in 1936 Rasmusen and the Farm Bureau also helped to educate farmers about the new federal Soil Conservation and Domestic Allotment Act, the successor program of the AAA that was designed to cut the production of cash crops and promote soil conservation through the raising of "soil-conserving" crops, such as legumes, grasses, and forage crops. Because of their efforts, DeKalb County became one of the first counties in Illinois to participate in the soil-conservation program.[26]

In September 1936, Rasmusen left the Farm Bureau to work for Tom Roberts at the DeKalb County Agricultural Association as a sales manager in its new hybrid corn department. His successor as farm advisor, Roy Johnson, was experienced in Farm Bureau work and had worked for a short time for the Illinois Agricultural Association before he took over the advisor's duties in December. It was a hard job made even more difficult by the Depression, and the period from 1936 through the end of World War II was a time of challenges for the Bureau. One major problem was simply keeping a farm advisor. Shortly after Johnson began, he resigned to enter military service, leaving the Bureau without a full-time, professional farm advisor for the first time in its history. Earl Wenzel, the part-time assistant, worked with the 4-H groups, helped with membership drives, and organized many of the Bureau's social events. A former 4-H participant, Ralph McKenzie, also worked part-time with the Bureau in 1936 and 1937, and Donald Mosher worked as a part-time employee through the summer of 1938. It was not until February 1939, that the Farm Bureau finally hired an official assistant farm advisor, Donald McAllister, who had the agricultural training and experience to assist farmers with soil testing and other scientific and technical issues.[27]

The Bureau struggled with finding and retaining a fully qualified farm advisor, but the most pressing problem was funding its programs. The Depression caused many farmers to quit the Farm Bureau because they could not afford the dues or were forced off the land altogether. At the same time, the Bureau was responsible for its usual services and attempting to expand its offerings to include a broad range of new social programs for hard-pressed county farm families. These social and professional services drained an already tight budget. Tom Roberts, the former farm advisor and then-secretary-manager of the DeKalb Agricultural Association, had been, and continued to be, an especially forceful voice for expanding the role of the Farm Bureau in the community during difficult economic times, but he and the Bureau leaders recognized that the economic crisis threatened the Bureau's ability to continue supporting important community programs such as 4-H clubs and the Rural Youth. For nearly twenty years, the Association and the Farm Bureau enjoyed a close relationship, and it seemed clear to county agricultural leaders that the future success of both organizations rested on their mutual support. Yet, by the autumn of 1936, the Bureau was having difficulty maintaining its programming, and prospects for improvement, even with the cooperation of the Association, were uncertain.

Late in the year, the Bureau and the Agricultural Association took measures to shore up their financial positions. One of the more success-ful joint ventures was the oil company. In 1931, the business had twenty-one delivery trucks serving DeKalb and neighboring Boone, Ogle, and McHenry counties. Looking for an alternative source of in-come to supplement membership dues and to make the petroleum com-pany more efficient, in early November 1936 DeKalb County Farm Bu-reau directors met with Farm Bureau representatives from Boone, McHenry, and Ogle counties to discuss purchasing a direct interest in the petroleum and allied businesses from the Agricultural Association. The Agricultural Association, dominated by men such as Roberts, Gunn, and Rasmusen, all of whom had close ties to the DeKalb County Farm Bureau, was eager to sell a portion of its oil business to the Bureau. The new cooperative association, financed with new preferred stock, was called 4-County Service Company. It lasted less than a year, when the Ogle County Farm Bureau elected to withdraw from the cooperative. The remaining three county farm bureaus re-formed their partnership into the Kishwaukee Service Company, in August 1937. The Agricultural Association, which had retained 50 percent of the total stock, named the manager and played the leading role in overseeing the business. A year later, in 1938, Kishwaukee Service Company opened a direct sales service station at the corner of Pine and North streets in DeKalb. The loyalty of Bureau members to their oil-delivery men remained a corner-stone of Kishwaukee Service Company success for several decades.[28]

Also in late November 1936, the Farm Bureau executive committee explored additional ways to further strengthen their institution's finan-cial position. Foremost in their minds were the tax advantages that would accrue to the Bureau if it could divest itself of its substantial pri-vate ownership interest in the DeKalb County Agricultural Association but still use income from the shares to generate funding for educational programming. More important, however, the Association's hybrid seed business was clearly poised to take off, and the Farm Bureau directors were interested in shielding their potentially valuable shares from the possibility that the Bureau would be forced to sell the stock to settle debts that the Bureau was incurring during the lean Depression years. "It should be the desire of every Institution," noted the president's sup-plemental report, "particularly a co-operative where the administrative offices are changed more often than they are in a strictly corporate form of organization, to throw every possible safeguard around its assets and to do whatever is in their power to assure perpetuation of the services that have built the organization."[29] The following month, the board of directors voted unanimously to grant eighty shares of DeKalb County Agricultural Association common stock that the Bureau owned to the Bureau's new, private, nonprofit corporation, the Rural Improvement Society. Directors of the Rural Improvement Society, drawn from the Farm Bureau, met twice a year and informed the Farm Bureau directors how much money it had accrued from the dividends on the Association stock to spend on "educational and other benefits to people within our county engaged in horticulture and agriculture in accordance with the requirements of our charter." In particular, the Rural Improvement Association regularly turned over funds "for the specific purpose of hiring a 4-H Club and Rural Youth leader and for defraying the costs in connection with similar activities being engaged in by the DeKalb County Farm Bureau."[30] Although these transactions provided a welcome security and stability to the Bureau's assets at the time, the stock transfer was perhaps most significant because it enabled the Bureau to take advantage of favorable tax laws for nonprofit corporations and shielded the core of the Bureau's equity in the DeKalb Agricultural Association from taxes and creditors at a time when the hybrid seed business was growing and showed promise of becoming an important component of midwestern agriculture. That action would, during the post–World War II period, lead to significant investment returns for the Bureau and make it one of the wealthiest and most influential in the nation.[31]

The closing years of the 1930s continued to be challenging for the Farm Bureau. Despite the difficult economic times, prospects for rural America, including DeKalb County, were better than they had been for some time. The Bureau maintained and even expanded its activities and programs. It promoted local forest preserves and tree planting. It assisted with local farm clubs, like the Sycamore Farmers Club, which predated the Soil Improvement Association but had fallen into a severe financial and leadership crisis during the Depression. The Bureau operated a tractor mechanics school that regularly offered tractor and farm safety advice to its members. In late 1935, the Bureau announced it would open a farm accounting school. In 1936, the Bureau started the Safe Drivers Club for young adults.[32]

In 1937, the Bureau began one of its most ambitious programs of the decade, a cold-storage locker service. Although DeKalb County had more electrical lines than most other Illinois counties, few farms had refrigerators, and fewer still had storage freezers, that could be used to reduce food costs and improve the quality of rural life. One solution to

the problem was a commercial cold-storage locker, and by the time the
DeKalb County Farm Bureau locker facility opened, there were already
around a hundred cold-storage lockers in Illinois. The DeKalb County
Locker Service, originally financed by a public stock offering, opened its
first cold-storage facility in Sycamore on May 1, 1937, and announced
in July that it was expanding to DeKalb. Each rented locker in the cooler
could hold as much as 275 pounds of cut meat. The locker service was
so popular that it had leased more than 100 lockers by the end of its first
year of operation. After two years in business, the Bureau owned an interest in five locker facilities run by the DeKalb County Locker Service,
Inc. The locker service, which eventually managed lockers in several
county communities, also provided butchering services and played an
important role in helping county families save money by offering the
means to slaughter and package farm-raised livestock economically.[33]

The Bureau also continued its political activities. It approved and supported, for example, the Illinois Agricultural Association's perennial efforts throughout the decade for tax relief on real property, fuels, and incomes. The most explosive local issue was the consolidation of the rural
schools. It was a controversial proposal that, among other things, would
eliminate many one-room elementary schools and rural secondary
schools. Rural school districts and families loved their small schools and
were jealous of their local control, but the expenses of running such
schools was high and the quality of education was inconsistent from
one school to the next. Consolidation advocates argued that large districts were more cost-effective when collecting and spending taxes, provided a broader range of educational opportunities, and offered county
children a greater variety of extracurricular activities than the small districts could afford. Reflecting both its core educational mission and its
tax reduction crusade, the Bureau leadership went on record as favoring
a consolidated system and sided with proponents who believed consolidation would cost less, improve education for rural children, and raise
the quality of life of farm families.[34]

The outbreak of war in Europe in September 1939 changed the world
situation dramatically. But, unlike World War I twenty-five years earlier,
the new war had little immediate impact on American agriculture. Although Britain and France spent lavishly on war material during the
months leading up to hostilities and in the opening months of the conflict, most of their urgent demand was for military hardware, including
airplanes, tanks, ships, and small arms. American factory employment
and wages increased, but nearly a decade of high urban unemployment

meant that few of those new jobs went to poor farmers or their children. As a result of the Depression, American industry had so much idle factory capacity, and American farmers had produced such surpluses, that new wartime manufacturing orders did not result in any real shortages or increased raw material demands that might drive up domestic commodity prices. Most ironic of all, in contrast to World War I, the most immediate effect of the new war was a decline of nearly 40 percent in American agricultural exports during the early stages of the conflict, as the succession of German victories across Europe cut off agricultural importers—like Denmark, Norway, France, and Belgium—from American trade.

Moreover, during the first winter of the war in Europe, the conflict seemed remote to most Americans, and there was little public interest in preparing the nation for direct participation in hostilities. Many Americans, in fact, were actively opposed to direct governmental action that might lead the country into another bloody, expensive, and nationally frustrating European war. Despite public hesitation, the Roosevelt administration believed that American involvement in the war was inevitable. It quietly and cautiously began planning for national defense and laying the groundwork for war preparation. The collapse of France, the exhaustion of British financial resources and evacuation of British forces from the continent, and the German invasion of the Soviet Union demonstrated Germany's threat to American national interests and accelerated the pace of the nation's war preparedness activities. Domestic industry quickly expanded production, but the agricultural sector responded more slowly. Moreover, throughout the Depression period, the emphasis of national agricultural policy had been to restrict production to stabilize prices through marketing and acreage controls. Federal efforts to maintain farm prices above market level had resulted in huge surpluses that were available to domestic industry and U.S. allies, and projected agricultural production was so favorable that few anticipated the sort of transition difficulties the nation experienced at the beginning of World War I. Policies rooted in Depression-era fears of commodity overproduction and low prices were difficult to set aside, and many farm-state politicians and agricultural administrators believed that supporting their rural constituents was as important as, if not more important than, preparing for defense. Nevertheless, farmers in DeKalb County in 1940 were in a better position economically than they had been in years. Statewide, 1940 was the fourth consecutive year of good crop yields, and prices for all but two of the fifteen most important agricultural commodities were higher than they had been in 1939. Average farm incomes in northwestern Illinois were up from about $1,600 in 1937 to more than $2,800 in 1940, reflecting an increase in earning per acre from $8.46 to $13.51 during those same four years. DeKalb farmers were even better off than many, with an average net income of $15.25 per acre in 1940. Average labor costs increased between 1938 and 1940, but they declined as a percentage of total farm income in the same period. The average total capital investment in DeKalb farms was just over $44,000, nearly $10,000 above the average in northwestern Illinois farms. Most DeKalb farmers were cautiously looking forward to even better conditions ahead.[35]

Their optimism is understandable. The United States undertook a series of policies designed to assist the Allies that encouraged production and provided financial benefits to American farmers. The Lend-Lease Act of 1940, for example, amounted to a federal government export program that shipped stored surpluses and new production to Europe. Government purchases of strategic commodities at guaranteed prices increased production and profits for certain critical agricultural products, such as eggs, poultry, dairy products, and hogs. The Agricultural Division of the National Defense Advisory Council worked to ensure parity prices relative to other goods and to prevent shortages of agricultural products. The Food Distribution Administration ensured that foreign demand did not result in shortages at home. In the spring of 1941, Congress passed a special price support legislation that helped to stimulate the production of basic agricultural commodities. Cereal grain prices, for example, increased about 50 percent.

At about the same time, the Anglo-American Food Committee and the Office of Agricultural Defense Relations set a broad agenda for agricultural assistance to Great Britain and laid the foundation for agricultural preparedness in the United States for the next several years. The U.S. Department of Agriculture encouraged the production of fats, dairy, meat, and tobacco to help meet British demand and offered subsidies to processors to improve methods of preserving and transporting these perishable products. By the autumn of 1941, prices for most farm commodities were higher than they had been since 1929, and governmental agricultural subsidies were so reduced from Depression levels that they accounted for only about 13 percent of farm incomes. In northwestern Illinois, average farm income jumped from $2,862 to $5,070, from $13.51 to $24.35 per acre, in just one year![36]

The indirect government involvement in agricultural affairs, and the reservations Americans had about participating in a foreign war, were swept aside by the Japanese attack on Pearl Harbor. The nation moved to full war mobilization, and the economic outlook for farmers improved even more dramatically. The rush of patriotism that accompanied America's entry into the war enabled the federal government to take aggressive actions in the agricultural marketplace that would have been impossible earlier. As the prospect for a greater wartime demand for food became real, the national emphasis shifted from restricting commodity production to encouraging farmers to grow as much as possible. Mobilization and the promise of government contracts and higher prices gave farmers the green light to accelerate agricultural production at rates much greater than peacetime production.[37]

The Roosevelt administration developed its agricultural policies based largely on the nation's World War I experience. Many of the president's advisors had firsthand knowledge with the earlier conflict; indeed, many held positions of responsibility and power during the first war. Consequently, the administration's agricultural policy makers shared fairly consistent ideas about how to manage agriculture during the new conflict. No one in the administration was interested in repeating the extensive social controls of the sort that had regimented rural America during 1917–1918. There was such public unanimity of support for the war after Pearl Harbor that there was little need to propagandize rural Americans to support the war effort, as had been done earlier. With these lessons in mind, the federal government sought to exercise as little control as possible over rural America. The few regulations that were put into place generally benefited agriculture. Government-imposed price controls, for example, applied to industrial goods that farmers needed but were not applied to agricultural products until a year later, in 1943. Farmers received special allocations of rationed goods, such as petroleum and tires. The government worked to ensure a steady supply of farm machinery, spare parts, and labor.

This light-handed management of rural America was motivated by more than past experience, however. Compared with World War I, farmers in 1941 were in a much better position to demand favorable treatment. One of the most important changes that occurred during the interwar years was that, increasingly, American farmers were joining together to express common interests at the national level. The American Farm Bureau Federation, originally founded in 1919, had emerged during the early 1930s as the most influential national representative for farmer interests. During the war, the AFBF monitored federal wartime farm activities, served as a liaison between the federal government and local farm bureaus, promoted new federal agricultural strategies, and offered its own policy initiatives. In 1942, for example, AFBF lobbyists persuaded Congress to grant draft deferments to all farm laborers producing "essential crops," a practice that often resulted in rural youth spending their war years out of military service and in the fields. It also succeeded in using its influence to block federal housing standards, minimum wages and maximum working hours, and collective bargaining for farm laborers. In parts of the rural South, the AFBF even persuaded the U.S. Department of Agriculture to threaten unwilling cotton hands with the draft if they left the fields to look for better work elsewhere. It lectured Washington politicians about the dangers of uncontrolled industrial wages and decried federal efforts to set ceiling prices for farm commodities. Above all, however, the AFBF feared that too much wartime government involvement in the agricultural sector would erode the farmers' cherished independence and worked to maintain agriculture as a key component of America's economy. "The situation is becoming very critical," wrote Clifford Townsend, "and it is necessary for every farmer to lend his effort in the struggle to maintain his position in the economic setup of this country. . . . Farm organizations are an important part of democracy in agriculture. Again and again they have taken the lead in formulating and improving the agricultural programs which have stood between farmers and disaster in times of crisis. . . . This is a challenge to farmers to join their own organizations and strengthen their own farm programs."[38]

Wartime demand for farm products and the emerging political power of rural America resulted, above all, in highly favorable federal commodity-pricing policies. Depression-era production acreage restrictions were lifted, but price supports remained. Those continuing New Deal price supports, in essence, dramatically reduced the financial risks of farming at the same time that farmers were freed to produce as much as possible. The result was soaring national farm prosperity. Not all rural folk shared equally in the abundance—the bottom quarter of farmers were hardly better off at the end of the war than at its beginning—but those who prospered did so handsomely. Although wartime profits were generally less dramatic than those farmers earned after the United States entered World War I, they were higher than anytime in the preceding decade. Nationwide, farm incomes rose by more than 150 percent between 1939 and 1944. In northwestern Illinois, average net farm income climbed to $6,162 in 1942—233 percent above 1939 net farm incomes in the region—before settling back to $3,749 in 1944, a figure that, though less than peak wartime incomes, was still about 142 percent above 1939 income and nearly 227 percent of 1937 levels.[39]

Unlike their parents, who spent their World War I profits on expensive land and equipment as if the good times would never end, farmers during the World War II used their profits to pay off mortgage loans and save for future investment. The Farm Bureau cautioned its members against rushing to expand their operations. "[I]f you are in debt, now is a good time pay out . . . if you are not in debt, this is a poor time to go into debt. . . . [We] strongly advise against tremendous inflation of land values such as followed the last war. [Farm Bureau] advises further that in this increased production program asked for by the government, that the farmer keep it on a temporary basis and not tremendously expand his plant facilities at high cost to himself and at tremendous risk of a loss following the emergency." Chester C. Davis, president of the Federal Reserve Bank of St. Louis, was even more direct in his advice to Bureau farm and home advisors. "Increased farm income," he said, "should be used to pay off debts and not as a basis for the expansion of debt. . . . The correct rule for the farmer is the same as for any other private citizen—pay off debt; spend nothing you do not have to spend. Farmers who are out of debt should use income to build up savings. Every farm family that can possibly set the money aside after reducing or paying off debts should invest, not in new buildings, not in more land, but in U.S. Saving Bonds." The Illinois Extension Service cautioned farmers to "plan for the final stages of the war and readjustment to postwar conditions. . . . Expansion should be on the basis of profitableness rather than on the basis of a hobby. In general, the build-up should be gradual."[40] Nationally, farmers took such advice seriously. Between 1940 and 1945, for example, farm mortgage debt decreased by $1 billion, or about 15 percent— a sharp contrast to the period from 1915 to 1920, during which farmers increased their debt nearly 70 percent, from $5 billion to $8.4 billion. By war's end, rural Americans also doubled their nonfarm equity, primarily bank deposits and government bonds, from about 10 percent of their total capital assets to 20 percent. At the same time, the total value of farm real estate jumped from $33.6 billion in 1940 to $56.5 billion by war's end. All of this new financial security was shared by roughly 25 million farmers, 5 million fewer than had been in the fields in 1940.[41]

In DeKalb County, as in the rest of the nation, the outbreak of World War II in Europe in 1939 had little impact on area farmers beyond increases in commodity prices. While the national government worked to improve military and civilian preparedness, the DeKalb County Farm Bureau continued offering its wide range of educational and social programs and services. In 1940, it branched out into another new direction.

After nearly twenty-eight years of emphasis on soils and crops, the Bureau concluded that the county's livestock producers deserved their own full-time advisor experienced in the special needs of cattle and swine production. It was a long overdue recognition that livestock was as important to DeKalb County as cereal crops. The Bureau hired Lee Mosher as its county organization director and livestock advisor on March 1, 1941, the first farm bureau livestock specialist in the nation. Mosher, a local Afton township farmer who had been active in Bureau activities since 1913, including 4-H and Rural Youth, was hired primarily to advise stockmen about marketing, fighting diseases, and raising better herds, but also to recruit members and assist generally with Bureau activities. The Bureau's timing in appointing Mosher proved to be very good, because the escalating demand for livestock products during the war made it vitally important that the county's herds were as productive as possible. He was especially active in helping stockmen treat and prevent disease in their animals and in herd management education.[42]

When the United States entered the war in late 1941, the Farm Bureau undertook a wide range of patriotic wartime community duties. The Bureau was active in community collection drives for scrap iron and rubber, civilian defense, bond drives, transportation pools, and food for defense and victory gardens. It offered advice to members about navigating wartime rationing regulations, price controls, and agricultural exemptions of all kinds to which farmers were entitled. Equipment maintenance was a high priority. "Men at the battlefronts on the farm," noted the *DeKalb County Farmer,* "are going to win or lose in the food-for-freedom campaign according to how ably they use their machines and keep them in repair. The farmers' weapons are tractors, plows, cultivators, mowers, seeding tools, harvesting equipment, trucks and wagons."[43] The Home Bureau selected as its theme for the 1942–1943 programs "Conserve for Victory" and designed a series of lessons that concentrated on "helping the homemaker do a better job in the present situation." As the war progressed, the Home Bureau turned its attention to broader national themes in programs such as "Brothers under the Skin," "Our Share in World Trade," and "Understanding the Peace Organization."[44] The 4-H clubs helped raise money for local and statewide military charities and even contributed money to purchase a fully equipped ambulance for the army. Illinois Rural Youth bond drives in 1944 helped pay for a medium bomber, christened "Illinois Rural Youth," and a North American P-51 fighter, named "Illinois Farm Youth." The DeKalb County Rural Youth also provided entertainment at fund-raising programs. The county advisor was a member of the county war board, the emergency wartime organization charged with implementing and managing a wide range of wartime regulations. One responsibility of the war board that was particularly important to the Farm Bureau and its members was to issue permits for new farm construction during a time of expanding agricultural production and severe restrictions on building materials.[45]

Perhaps the most chronic difficulty that the Farm Bureau helped solve during the war, however, was a shortage of farm labor. After nearly a decade of labor surplus in the agricultural sector, the armed services and industry absorbed labor at the very time when the war demanded higher levels of farm production. Although the local selective service board liberally granted 2-C agricultural deferments from military service to area men to operate their farms, there were still too few hired men to go around. The war board and the Farm Bureau worked closely with the selective service board in DeKalb and neighboring counties to retain agricultural workers and allocate extra labor at critical times, such as harvest. William R. Randall, farm labor assistant, arranged for thousands of extra laborers in 1945 to assist not only area farms, but also the DeKalb Agricultural Association, the Sycamore Preserve Works, and the War Hemp Industries mills. During the war, even German prisoners of war earned money by helping local farmers with their harvests. It was a difficult task that was never entirely resolved, and most area farm families made up the labor shortfall themselves simply by working longer hours.[46]

The wartime labor shortfall also caused disruptions in the Farm Bureau at the very time that the Bureau needed to have an advisor and several assistants to make certain they had adequate help to serve the membership. Because of a wartime shortage of qualified farm advisors, the Farm Bureau experienced a high turnover rate in that position and was forced to hire a series of farm advisors and assistants between 1939 and 1946. Roy Johnson, who was originally hired as farm advisor in late 1936 but resigned to meet his military obligations, returned to become farm advisor beginning December 1, 1941. Within a few weeks, however, Johnson was recalled to active service and commissioned a lieutenant in the air corps. Assistant advisor Donald McAllister, hired originally in 1939, succeeded Johnson as farm advisor in 1942 and remained through early 1944. The Bureau hired Carroll Mummert as assistant farm advisor in 1943 and promoted him to farm advisor in 1944 to succeed McAllister. The Bureau also hired Wilbur Goeke as assistant for two years between 1941 and 1943, special program assistant Harvey

Schweitzer in 1944, and assistant advisor Dale Philips for six months in 1945. On January 1, 1946, the Bureau hired Ward Cross as assistant farm advisor. In February, 1943, the Bureau hired Bernice Engleking to replace Helen Johnson as home advisor.[47]

Despite the problem with keeping advisors in DeKalb, the Bureau succeeded in maintaining continuity in its educational and service programs. The overall Farm Bureau and Extension Service wartime program in the county, known as the "Wartime Educational Program," was a continuation and intensification of traditional Bureau services. Area farmers, for example, relied on the Bureau and advisors to prescribe and deliver soil additives and to advise on crop-rotation patterns, disease and pest management, and new crops—one of the most difficult assignments for the farm advisor during 1943 was assisting area farmers grow 10,000 acres of hemp to replace jute, which was in short supply because of the war. The federal government built three hemp-processing plants in the county, in Shabbona, Kirkland, and Genoa, but none of the growers had any experience with hemp, about 10 percent of the crop failed, and while the returns were good, most farmers quickly returned to more familiar crops.[48] Thanks in large measure to the excellent programming and assistance of the Farm Bureau, DeKalb County was so efficient and effective in food production that the federal War Food Administration cited the county as one of only three in Illinois to receive the administration's "A Award" in 1943. "Overcoming great difficulties by decisive action, laboring with determined devotion, joining together in making wise use of all their resources," read the citation, "the farm people of this county have answered their country's call for utmost food production. They have thus contributed in fullest measure to the cause of the United States of America and to the preservation of human freedom." Henry Parke, one of founders of the Soil Improvement Association, accepted the award on behalf of the DeKalb County farmers.[49]

The Bureau also worked during the war years to improve its member cooperative services. In 1943, for example, the Bureau ended its practice of providing livestock vaccines directly to members and incorporated DeKalb County Producers Supply to undertake the job. Producers Supply quickly expanded its offerings to include serums, pesticides, medicines, and veterinary tools such as syringes. The Bureau-affiliated Kishwaukee Service Company worked with federal regulators to provide gasoline and other rationed petroleum and rubber products to area farmers. It also advised farmers about the most efficient use of those products.[50] Demand for the Bureau's locker service was so great that the Bureau built new slaughter facilities, improved its smoking and curing capacity, upgraded its lard-rendering equipment, and built new storage lockers in each of its six plants. In the summer of 1945, the Bureau worked closely with the Northern Illinois Breeding Cooperative for the artificial insemination of dairy cattle to improve production herds in the county. At the end of the war, the Bureau organized a veterans advisory committee to help returning servicemen adjust to civilian life.

When the war ended in 1945, DeKalb county residents rejoiced in the Allied victory, welcomed those who returned home, and mourned those who did not. The wartime demand for agricultural products boosted commodity prices and the general prosperity of the county. Because of wartime shortages, however, most people found very little to buy with their new incomes. Many saved their profits and others paid off debts, but nearly everybody was relieved that the economically troubled decades of the 1920s and 1930s were past. For the Farm Bureau, which had worked so hard for so many years to help farmers weather the Depression and then meet the agricultural demands of the war, the future, for once, looked promising. Yet, despite the optimism, the Bureau also faced a future full of uncertainty. At the end of World War I, the agricultural sector slipped first into recession and then into a prolonged economic slump that ruined farmers and agricultural businesses throughout the county; it was a hard practical lesson in market economics no one in the county wanted to repeat. Although farm prices were higher during the war, wartime price controls had dampened agricultural income in comparison with higher industrial wages and consumer prices, a situation that hinted at the sort of economic imbalance that worked such despair on farmers during the 1920s. Postwar inflation, fueled by pent-up demand and savings, loomed on the horizon. Agricultural production soared, but costs for the specialized seed, soil additives, machinery, and other improvements required to sustain that level of production were also rising. Above all, DeKalb County farmers faced a challenging new era of farming in which the pace of technological change and the structure of agricultural markets would require them to become even more studious, efficient, and financially astute in order to succeed. The Farm Bureau would continue to be a critical source of educational and service support to help them face the promise, and uncertainties, of their postwar future.

5 POSTWAR YEARS

THE FARM BUREAU AT HIGH TIDE

•   On May 8, 1939, the DeKalb County Farm Bureau hosted the "Past Presidents Banquet" to honor its pioneer leaders on the occasion of the twenty-fifth anniversary of the Smith-Lever Act. Two of the Bureau's past presidents were already dead, but L. D. Sears, Aaron Plapp, Henry White, and Henry Parke were on hand to accept warm tributes from grateful Farm Bureau members. E. E. Houghtby was the master of ceremony, Mrs. Roy Tindall played a piano recital at the gathering, and the Franks Community Male Quartet sang. The evening's soloist was Mrs. Earl Wenzel. Former farm advisors Eckhardt, Roberts, and Rasmusen offered remarks on the history of the Bureau and highlighted each leader's contributions to the organization. The Bureau also presented a watch to each of them to commemorate the occasion. Although it was a local celebration, important regional and national agricultural personalities of the era attended the gala event—one reminder of how influential the DeKalb County Farm Bureau had become in a quarter century. L. D. Sears best summed up the emotions of the honorees in a thank-you letter to the party's sponsors: "There are certain events in a man's life that stand out as high spots. To me the Past Presidents Banquet of the Farm Bureau was one of those events. While I was president of the Farm Bureau, I did not consider that I was doing anything more than any one else would have done—the best he knew how under all the varied circumstances."[1]

The guests at the event understood and acknowledged the contributions these men made to the history of agriculture in DeKalb County. Yet no one fully appreciated that the celebration came at a watershed time for the DeKalb County Farm Bureau. While the guests toasted the past and celebrated the Bureau's success, a new era in agriculture, one that was full of trials both novel and familiar, was rapidly approaching. International events, especially, were already shaping a future that would be vastly different than anyone at the banquet could imagine. Half a world away, Japan was already deep into its conquest of Asia; a few months after the celebration, World War II erupted in Europe. Two years later, America itself was plunged into the carnage. Domestic agriculture underwent significant changes during the war, and old ways of farming, and administering the Farm Bureau, were swept away by the onrush of modernity. The postwar world was full of challenges for DeKalb farmers, and they looked to the Farm Bureau, as they had done for more than three decades, to stand with them as they faced the future.

Most farmers in America, and especially those in DeKalb County, emerged from World War II with greater financial security than they had at the end of their fathers' war. Nationwide, net farm income rose from about $16.5 billion in 1941 to $21.2 billion in 1944. Total cash income to Illinois farmers in 1944 was 120 percent more than the cumulative basis of the years 1935 to 1939, and nearly 440 percent more than Illinois farmers were earning just twelve years earlier, in 1932.[2] Reflecting upon their previous experience during the First World War, most in the farming sector of the nation's economy worried that they would suffer a similar postwar economic bust and recession that had staggered American agriculture in the 1920s. Their worst fears about financial disaster, however, did not materialize. As happened after World War I, some of the wartime agricultural profits went into farm machinery, new land, and technological improvements. In general, however, wisdom gleaned from past experience made farmers more cautious. They were careful not to overextend themselves, speculate recklessly in land and animals, plunge deeply into debt, or assume that the flush times would last. The national economic climate was also different. Postwar European relief programs, such as the Marshall Plan, absorbed some of the increased wartime agricultural production and distributed large quantities of American food to war-ravaged countries. More significantly, domestic consumer buying burst forth once rationing and other restrictions ended. Fueled by the incomes earned and saved during the conflict, surging domestic consumption triggered a brief spike in commodity prices. The resulting inflationary cycle benefited farmers in the short run. National farm policies rooted in Great Depression agricultural programs and the new welfare-state political philosophy that emerged during the New Deal shielded farmers from economic stagnation and helped cushion the transition from war to peacetime economy. A strong nonfarm economy also meant that the overall national prosperity was far greater than it had been during the Depression years.

Increased agricultural production combined with favorable government regulations generated a level of rural prosperity and security unseen in rural America for nearly two decades.

One feature of this new, postwar farming economy was a productivity revolution that introduced improved varieties of crops and animals. During the 1920s and 1930s, plant geneticists, like Charlie Gunn and Thomas Roberts at the DeKalb County Agricultural Association, discovered that if they inbred, first, two—and, later four—strains of corn, they could produce plants that displayed more favorable characteristics than any of the original strains or the best open-pollinated plants. Plant scientists were especially attracted to corn because it is fairly simple genetically, its fertilization process is easy to control, and it grows well under the right conditions and produces a readily marketable crop. Hybrid varieties first appeared in the mid-1920s, but farmers who were reluctant to change their traditional farming practices, especially during the times of dire financial crisis, were slow to accept the new hybrids. Moreover, although hybrids boosted crop yields for farmers, the best hybrid characteristics lasted only one growing season, forcing farmers to purchase new seed from the hybrid seed companies each year. That made hybrid seed more expensive to plant than open pollinated varieties. Successful breeding programs eventually produced hybrid corn in marketable quantities at reasonable prices that won over even the most skeptical farmers. During the 1930s, hybrid corn became increasingly popular in the corn belt, including DeKalb County, in part because it enabled farmers participating in the Agricultural Adjustment Act acreage restrictions to maintain high yields on smaller parcels of land. In 1935, American farmers planted 2,500 acres with DeKalb County Agricultural Association hybrid seed; in 1940, they planted more than 4 million acres in DeKalb hybrids. Demand for hybrid seed increased markedly during World War II until, by the end of the war, nearly all corn raised in DeKalb County and throughout the corn belt was hybrid. Genetic scientists achieved similar results in other crops and eventually applied new techniques of selective breeding to improve poultry and livestock.[3]

Hybrids and improved livestock increased farm output dramatically. Corn production, for example, increased nearly three-fold during the three decades after the end of World War II, while cropland devoted to corn remained relatively constant. Practical application of the new genetics technology would not have been possible, however, without the mechanization of the modern American farm. The most important new tool was the tractor. Although draft animals continued to play a role in local farm life, most county farmers had already made the change to machine power well before the war began. By the end of the war, tractors were the primary source of power on 40 percent of American farms; by the end of the 1980s, there were more than two tractors per farm. Disk plows, planters, and a host of new machines made farms more efficient. Mechanical harvesting machines were not new—the Marsh Harvester Company in DeKalb County first produced horse-drawn versions of them in the 1850s—but by the mid-twentieth century harvester combines were self-propelled, complex, and indispensable vehicles capable of stripping huge fields in a short time. Electricity also encouraged mechanization and improved the lifestyles of rural Americans.

The third element of the productivity revolution was chemicals, especially petroleum-based compounds. Gas and oil for the new farm machines was already in great demand in rural America by the 1920s, and the Kishwaukee Service Company, which was affiliated with the Farm Bureau and the DeKalb County Agricultural Association, owed its immediate success in the late 1920s to regional farmers making the transition to mechanized agriculture. Scientific advances in petroleum chemistry during the 1930s and 1940s produced a wide range of new petroleum-based products for the farm, including high-octane gasoline, diesel fuel, improved high-temperature lubricants, synthetic rubber, and plastics. Chemists also developed new fertilizers and pesticides that were much more potent than what was available to farmers before the war. Nitrogen fertilizers were especially important for corn and other nitrogen-hungry crops. One of the most prominent of these new chemical fertilizers was anhydrous ammonia, a gas that converts to a liquid in contact with air, which became widely available shortly after the war. Farmers in DeKalb County could now devote their fields to two crops, principally corn and soybeans, and ignore the complex, time-consuming, and often expensive crop-rotation and manure-additive schemes that county advisors, like William Eckhardt, had prescribed to meet the soil preservation demands of an earlier era. Another important chemical that appeared after the war was dichlorodiphenyltrichloroethane, or "DDT," the first successful modern insecticide available to farmers. Although farmers had used insecticides for decades, none was as potent and apparently safe as DDT for a broad range of crop and food applications. A new kind of broad-leaf growth hormone herbicide, 2,4-dichlorophenoxyacetic, or "2,4-D" as it is commonly called, enabled farmers to eliminate weeds chemically. Soon, companies introduced a range of other herbicides that were effective on a broad range of weed species. By ridding farm fields of weeds, farmers cultivated less, saved money in labor and fuel costs, and produced higher yields. New defoliants stripped plants of their leaves, making mechanical harvesting easier and more cost effective. Antibiotics and growth hormones rid livestock of traditional bacteriological diseases, accelerated weight gain in animals, and improved dairy production. The modern cornucopia of new chemicals promised to usher in a new, chemically enhanced era in farm prosperity.

The sweeping changes brought by these technological advances affected some agricultural areas more than others, but every successful farm in the postwar era relied on the new products to boost production. DeKalb County was no exception. Throughout the country, farmers increasingly substituted low-cost fossil fuels, in the form of gasoline and diesel, chemical fertilizers, electricity, and farm machinery for expensive and less efficient human and animal labor. As a consequence, the energy output of American agriculture, measured in terms of total calories of energy output compared with energy input, declined. In economic terms, however, the new technologies enabled farmers to dramatically increase their efficiency and profits as long as fossil fuels were comparatively inexpensive. Many of the changes were already taking place as early as the 1920s, especially the transition to mechanized farming, but the pace of change during the postwar years was breathtaking. At the same time, the consistency of food improved, losses to pests and disease were reduced, yields became more dependable, and farmers were more secure about the success of their crops than at any other time in the nation's history. Prosperous farms became sophisticated business operations that demanded expertise to a greater degree than even the most prescient farmers of the earlier decades could have foreseen.[4]

Yet, despite all these new developments, farmers in the postwar period faced a host of familiar challenges. The most pressing one was excess farm capacity, which resulted in financial strains when high demand for farm products declined, resulting in net farm incomes that did not match the pace of industrial wages or overall national prosperity. In 1941, annual net farm income nationally averaged about $2,000, and industrial income was about $2,400. In 1944, farm income was close to industrial wages, each about $3,200. By 1958, however, farm incomes had eroded to $2,120 compared with industrial wages of $3,500. As was the case during much of the interwar period, farmers in the 1950s and 1960s produced at high rates but the domestic and international markets did not expand to match supply. The problem was temporarily relieved during the war years and the immediate postwar period, when the Allies and starving nations soaked up American's excess production, but it was not a permanent solution. International demand fell in the late 1940s, when those nations recovered their own agricultural production and economic health, leaving an excess domestic output that held down commodity prices. Farm production also exceeded domestic population growth. Between 1940 and 1960, the national population increased on average about 1.7 percent annually, while farm production increased on average about 2.5 percent annually.

Domestic consumers changed their buying habits, placing downward pressure on farm incomes. Large grocery stores offered American consumers a broad range of new agricultural products, many of which came from abroad, where labor and production costs were lower than what American farmers paid. Consumers developed more discriminating tastes and became more thoughtful buyers who responded quickly to price changes at the retail level, shifting to alternative foods when prices fluctuated upward. New home refrigerators and freezers gave consumers the ability to store foods for longer periods and take advantage of favorable short-term price fluctuations. High domestic agricultural production and importation of commodities translated into lower food costs for American consumers, who devoted more than a third of their incomes to food in 1945, but less than 25 percent in 1970, and only about 16 percent by the end of the century. Consumer clothing tastes changed, and synthetic materials competed successfully with natural fibers. As the twentieth century wore on, natural fiber clothing was increasingly manufactured abroad from foreign-grown cotton, linen, and wool. Consequently, during the closing years of the 1950s, the U.S. Department of Agriculture estimated that the farming sector had excess production of between 6 and 8 percent annually. Neither overseas buyers nor federal policy measures, like international famine relief or homeland poverty programs, were large enough to make much dent in the surplus and boost sluggish farm incomes. At the same time, costs for farm necessities increased relentlessly, squeezing farmers between rising costs and stable prices. Farm costs increased 19 percent between 1947 and 1959, but the value of all farm products sold declined 11 percent. Buffeted by these forces, farmers kept afloat financially only by constantly improving farm efficiency.[5]

The new technological achievements that enabled American farmers to increase domestic production were also helping chronically impoverished and undernourished nations become self-sufficient in food, further dampening foreign demand for American products. The strong dollar and heavy European subsidization of their own agricultural sector, assistance that far exceeded domestic price supports, made American products expensive overseas while encouraging European competition on other areas of the world. Economies of scale achieved by large, consolidated owner-operator farms, which industrialized agriculture more than ever before, put further downward pressure on family farm incomes. Between 1945 and 1970, the consumer-price index increased by 116 percent, but the total value of farm output rose only about 40 percent. During that same period, the already high capital investment in an average American farm steadily increased as farmers attempted to compensate for sluggish returns by improving their efficiency and yields. Consequently, in 1945 a farm returned about 17 percent of its value as income, but by 1970 farmers received only about 6 percent of their total value as income.

These economic problems resulting from high production during the postwar decades are also attributable to several social and market assumptions that were deeply ingrained in the American mind. Reliance on scientific advancements to resolve complex social and institutional problems, one of the most powerful secular faiths of the twentieth century, meant Americans, especially farmers, eagerly embraced new technologies as desirable solutions to their problems, and farm capacity surged while prices fell. Market conditions, too, offered a powerful incentive for farmers to employ the new techniques. Despite traditional farmer rhetoric about the benefits of free-market capitalism, few farmers were in any position to participate effectively in the marketplace—as often as not, the unbridled operation of agricultural markets caused severe hardship for American farmers. In reality, the market for farm products has historically been inelastic, so incremental increases in output drive agricultural prices down. In the late 1950s, for example, the elasticity factor in agriculture was negative .25.[6] In other words, a 1 percent increase in supply brought a 4 percent reduction in prices. Consequently, in the agricultural marketplace, individual farmers who had little influence over markets were always looking for the best means to cut cost per unit of output, and technological advances often provided the most obvious means to accomplish that. The inelastic demand is exacerbated by an equally inelastic supply of farm products. Farmers—committed to long-term fixed costs, confronted by production cycles measured in growing seasons, and tied to a capital base that is difficult to convert into alternative profitable uses—find themselves in no position to easily adjust their production to meet fluctuating market conditions. It is likewise difficult, if not impossible, for most farmers to stop production for any length of time, hoping for better prices in the future, and expect to remain ahead of their creditors. Only the intervention by the federal government kept farm prices stable during the postwar period. Federal subsidy programs that supported commodity prices kept farm incomes generally constant from the early 1950s through the late 1960s—livestock prices, which were not subsidized to the same degree, fell dramatically during the same period.

Another familiar condition during the postwar years was the steady decline in the number of people working the land. At the end of World War II, there were about the same number of farms in America as there had been at the eve of World War I, about 5.8 million. Thirty years later, farm consolidation meant that there were only about 2.85 million farms. One principal cause for the decline was the flight of farmers from rural America to the cities, which began in earnest during 1920s and accelerated during the war and in the following decades. In 1930, there were about 30.5 million people living on farms, around 25 percent of the national population. Ten years later, the same number accounted for 23 percent of the population. A steady outflow of America's rural population—at a rate of about 900,000 each year between 1940 and 1960—meant that only 25 million farmers remained on the land in 1950, about 17 percent; 22.4 million, or 14 percent, in 1955; and only 21.2 million, 12 percent, at the end of the decade.[7] By the end of the twentieth century, farmers accounted for less than 5 percent of the national population.

As is the case with any migration, there are factors that acted to both push and pull people away from their rural homes. Many farmers were forced out by agricultural market changes. Others left when mechanization erased farm labor jobs. Still others simply gave up farming as too difficult in the modern era, sold out to their neighbors or corporations, and took jobs in towns and cities. The general national prosperity of the postwar period easily absorbed these urban newcomers. The lure of urban excitement and the perceived limitations of farm life also drew young men and women away from rural America. Young people and retired farmers found that they enjoyed the conveniences of town living. The outflow of rural people strained local communities and undermined social institutions. In some places, it accentuated class differences between successful and marginal farmers, as well as between rural and urban folk. Falling rural populations also meant farmers wielded less political power at all levels. At the same time, however, the migration out of the countryside also helped those who remained behind because it meant that there were fewer people competing for farm income.

Finally, although farm life in America during the first half of the twentieth century was improving, in many respects it remained hauntingly familiar to what it had been for generations past. The postwar agricultural transformation, however, accelerated the pace of change in rural living. A middle-class lifestyle, equivalent to living standards of city folk, had been a dream of successful farmers since the nineteenth century, but it became a standard expectation for farmers after World War II. Consolidated rural childhood education, telephones, radio, and finally television, which became widely available during the 1950s, swept away the historical isolation and parochialism of American farmers. "Radio and television," said Farm Bureau founder Henry Parke in 1954, "are the most potent forces of education the world had ever seen. These two reach into the most remote corners of the world to stamp out ignorance. . . . Through radio and television, we can hope to reestablish the underfed nations of the world. Through them, we can christianize the heathen and establish world peace. . . . In the electrical field, we find vast areas of progress here and now. We can but marvel at what is ahead of us in this field." Cars, roads, and bridges compressed distance and increased opportunities for social and commercial interaction. Modern medicine improved the health of rural folk and rid the countryside of once crippling and deadly diseases like scarlet fever, measles, polio, and mumps. Pesticides controlled insects and vermin. Changes in farm sanitation also reduced diseases. Outhouses gave way to sanitary indoor plumbing and septic systems. In the early twentieth century, most farm wells in DeKalb County were only twenty to thirty feet deep and were easily contaminated by farm runoff, but by midcentury wells were bored 150 feet or deeper into rock aquifers and sealed off from farm wastes.[8]

Many young men and women who served in the military or attended college returned to the farms with new ambition to enrich rural life, which looked plain in comparison to their worldly experiences. The role of women on farms also changed. They had fewer household responsibilities because they had fewer children, and the new machinery meant that farm men hired less farm labor. Laborsaving electrical home appliances took some of the drudgery out of housework. Increasing numbers of rural women took jobs in small towns or nearby cities. Skills like canning, butchering, soap making, and sewing, which had once been necessities of country life, became hobbies as farm families increasingly turned to stores for commercially manufactured products. As farms moved away from diversified agriculture and more toward to specialization, traditional women's farm chores, such as raising poultry, disappeared. At the same time, the mechanization of agriculture made farm jobs easier and enabled women to engage in traditionally male chores, further reducing the need for hired labor. Farm children experienced equally dramatic changes in their lives. During the postwar period, fewer children worked regularly in the fields. Farm work became less important as a practical educational experience for children. The 4-H, once an indispensable steppingstone for preparing rural children to become farmers and farm wives, broadened its offerings to include topics that were only distantly related to farm work and welcomed urban children who had no opportunity or expectation of working the land. By the end of the twentieth century, most 4-H participants lived in towns and cities. Rural parents expected their children to complete high school. Many young adults went to college to prepare themselves for their lives as modern farmers or, as farming opportunities shrank, to enter the workforce outside agriculture.[9]

At the end of the war, farmers in DeKalb County experienced many of these economic and demographic changes. They were, on the whole, better off financially than they had been at any time in the preceding twenty years and were among the most successful farmers in America— DeKalb County residents commonly assumed, with much justification, that their community was the best-fed in the nation. In 1945, there were 2,224 farms in the county, totaling about 384,300 acres, with an average size of about 179 acres. A number of farms were between 500 and 1,000 acres. The county ranked sixth in the state in corn production, with average yields of about sixty-five bushels per acre; average oat yield was about seventy bushels, and soybeans averaged thirty bushels. DeKalb County farmers annually produced 90 million pounds of milk, 22 million pounds of pork, 100 million pounds of beef, and 3.5 million pounds of mutton—enough output to feed a city of a half million for a year! More important, in 1946 DeKalb farmers earned an average net return of $38.75 per acre and an average return on farm capital investment of nearly 21 percent. Livestock returned 162 percent of feed costs at slaughter, and the average return for dairy was nearly 270 percent of feed costs. County farmers planted 6,700 acres to produce hybrid seed and another 19,100 acres in canning crops such as sweet corn, peas, and lima beans. Annual labor costs were just over $19.00 per acre, and machinery costs were $13.00. Land values were between $200 and $250 per acre, with some prime land selling for $300. Those prices, considerably above normal valuations of between $125 and $150, demonstrated the demand for DeKalb County land and the optimistic outlook that buyers had about farming in such a prosperous area.[10]

Such confidence was well founded. Wartime prosperity made DeKalb County a more appealing place to live for those who remained on the farm; a variety of local industries and businesses provided opportunity to those who chose to work for wages. The Farm Bureau provided important services to its members during the war to maximize their production and maintain profits, and the dramatic shifts in agriculture caused burgeoning demand for a wide range of Bureau services at war's end. The Bureau seamlessly shifted its attention from war production to peacetime agricultural success. The scale of Farm Bureau activities also increased, demonstrating not only the willingness of the board of directors to undertake expansion to address new challenges, but also the financial maturity and institutional confidence to invest vast sums of money and energy in the county's future.

Traditional Bureau services flourished. The Bureau's insurance program surged after the war, with thousands of policies in force for protecting area residents from fire, wind, hail, and personal accidents. Life insurance and employer liability policies were also popular. In 1950, insurance brokers even offered polio insurance for the first time, with 700 policies in force within four years.[11] Auto insurance, however, was the overwhelming favorite for Farm Bureau members, especially once wartime rationing ended and area residents were able once again to buy new cars and drive freely. In addition to the savings farmers enjoyed from participating in the affiliated insurance programs, by 1945, the dividends paid to policyholders in DeKalb County from all the insurance services exceeded the total cost to DeKalb farmers of operating the DeKalb County Farm Bureau. The Bureau's DeKalb County Producers Supply Company, which handled hog cholera serum, inoculation virus, and veterinary care hardware, sold more than a million milliliters of inoculation drugs and experienced an increase in demand for products of about 20 percent in the years immediately following the war. Postwar demand for petroleum products and chemical fertilizers from the Kishwaukee Service Company surged ahead—it sold around two million gallons of gasoline and close to a million gallons of other fuels annually during the first years after the war. Dividends Kishwaukee Service paid to Farm Bureau members increased from $27,000 in 1944 to $44,720 in 1946. The Locker Service Company expanded during the war years to include six plants with a total net worth in 1945 of $91,700. In 1949, the Bureau sold sufficient preferred Locker Service stock to enable the company to build a new state-of-the-art slaughtering, rendering, and pork-curing plant outside DeKalb to meet the countywide demand for butchering. "No longer is it necessary," ran one advertisement for the new facility, "for DeKalb County farmers to pay high prices for unsanitary on-the-farm butchering." During its first year of operation, the facility packed 1,700 head of hogs, beef, and lambs for Locker Service patrons. By the middle of the decade, the plant was also cleaning a thousand chickens a week and offering for sale to its members fresh, canned, and frozen fruits and vegetables. Also in 1949, the Bureau organized the DeKalb County Grain Company, a cooperative to market grains and purchase livestock feed, grass and legume seed, fertilizer, and fencing materials. A year later the grain company, with its main elevator in Somonauk, had a net worth of $72,500.[12]

The Bureau's commitment to educational programs and services was equally strong in the immediate postwar years. The Bureau helped organize the DeKalb County Soil Conservation District in 1946 and offered it moral and financial support. The Bureau's livestock specialist, Lee Mosher, assisted hundreds of DeKalb stockmen with their marketing programs. In 1948, when the DeKalb County Farm Bureau's board of directors authorized formation of a tax research department, it became the first farm bureau in the state to initiate an active program aimed at a more economical and efficient local government. The goal of this new division was to assess the effectiveness of local governments, courts, public services, and elected and appointed officials and make recommendations to Farm Bureau members about governmental policy changes.

The Home Bureau, heavily supported by the Farm Bureau and under the leadership of Bernice Engleking and her assistant Marbry Fay, and later assistants Jean Neese Picket and Dorothy Giese, had more than 500 members in 19 local units, and expanded to 22 units by 1950. Women completed projects on a broad variety of topics, including "Desserts for Entertaining," "Table Decorations," "Growing Old Gracefully," "Helping the Child Develop Emotionally," "Selection of Clothing Accessories," and "Use of DDT in the Home." They also worked to support community polio, tuberculosis, and Red Cross drives. The 4-H youth programs, which the Bureau also subsidized, had a fairly stable participation of around 450–500 members, including around 350 girls in the "home economics" section and a handful girls in the agricultural section. The easing of wartime restrictions and the flood of returning servicemen generated renewed interest in Farm Bureau–sponsored social programs, the most popular of which was Rural Youth. During the war the future of "RY" seemed in doubt, but the organization rebounded in

1946 and organized a seemingly endless cycle of meetings, seasonal parties, formals, box socials, hayrides, talent shows, and other recreational activities.[13]

The Bureau, through its association with state and federal extension services, offered area farmers other important farm services. More than a hundred farmers annually participated in Farm Bureau Farm Management service, which provided farm analysis and management programming, gave away farm account books, assisted with filing tax returns, and offered general financial advice to members. The growing interest in the revolutionary new farming practices generated interest in DDT and other new pesticides that required instruction in insect threats and effective application. Mechanical and chemical weed-control education programs introduced farmers to the latest developments, including the proper use of 2,4-D and other herbicides. The most common service that farmers demanded, however, was sophisticated soil testing, which the Bureau helped provide through its association with the state and federal extension services. The Bureau built a soil laboratory in 1945, and, by 1950, its soil technicians, like Fred Brock, provided analysis for more than 33,000 acres of land, prescribed fertilizer requirements, and offered long-term soil fertility and maintenance recommendations.

The importance of Farm Bureau to farmers in the years after the war is reflected in a substantial increase in membership. In 1946, membership stood at 2,250; in 1949 there were 2,900 members; and there were 3,340 members in 1954. In addition to the new membership fees and greater participation in the Bureau's associated companies by the members, Bureau finances during and after the war improved significantly. The most important source of revenue was the Bureau's DeKalb County Agricultural Association stock, the core of which the Bureau originally acquired in 1918 and which, since 1937, was held by the nonprofit Rural Improvement Society to support Bureau activities. The DeKalb County Agricultural Association's successful production of hybrid corn boosted its profits, and stock dividends, exponentially. In 1930, when hybrid corn was still in fierce competition with open-pollinated varieties, shareholder equity in "the AG" was slightly over $141,000; in 1940 it was about $1.95 million. By 1949, however, shareholder equity had grown to an astonishing $14.1 million. "The AG" also expanded into the oil development business, further boosting its capital returns. Dividends from its stock provided a steady, secure, and increasing source of income for Farm Bureau projects.[14]

The surging membership and demand for Bureau services, however, underscored the fact that existing facilities were woefully inadequate for the new era. The Soil Improvement Association opened its first office on North Third Street, in DeKalb, in 1912, and remained there until it bought the old North School in DeKalb in 1918. Over the next thirty years, the Bureau made many improvements to that wood-frame building, including adding a new floor and grain elevators and bins to store crop seed. As early as 1935, however, expanded services made it apparent to the Bureau's board of directors that even with the improvements to the building, it was too small to house the Bureau, offices of its associated companies, the farm advisor, the Home Bureau, and all the various other governmental tenants, such as the Federal Farm Bank, the Agricultural Adjustment Act Administration, and the Extension Service. In 1935, the board of directors began to set aside money and make plans for a new building.

The Depression and war delayed construction, but in 1946, the Bureau appointed a building committee and bought land in the city of DeKalb for its new home.[15] For the next four years, the Bureau considered building designs. As the plan evolved, it became clear that the original parcel would be too small, so the Bureau bought additional land to accommodate the increasingly ambitious architectural design. In the spring of 1950, construction began and was completed a year later, at a cost of approximately $200,000. The building was a substantial improvement over the "old school house." In addition to providing modern office space for the Bureau and its tenants, the fireproof brick and steel building housed a large auditorium for Bureau functions and community activities and a state-of-the-art soil-testing laboratory. Bureau board members were justifiably proud of their new home. "We are determined," wrote E. E. Houghtby, chairman of the building committee, "that this building should stand for generations as a great memorial not only to the pioneering work of our forefathers but as a fitting reward for the unselfish service which has been rendered by DeKalb County leaders in the Farm Bureau organization from its beginning in 1912 until the present time." Parke, the scion of the Bureau, referred to the new building simply as a "temple of agriculture."[16] The new building became, in fact, a one-stop agricultural emporium housing the offices of the Farm Bureau, DeKalb County Grain Company, Home Bureau, DeKalb County Locker Service, Illinois Agricultural Association financial assistance services, Kishwaukee Service Company, National Farm Loan Association, Northern Illinois Breeding Co-Op, Production and Marketing Administration, and the Soil Conservation Service.[17]

By the early 1950s, it was already evident that the DeKalb County Farm Bureau was the wealthiest and most successful bureau in the state and among the leading bureaus nationally—its new building was a tangible confirmation of that success. Yet for all its achievements and wealth, the Bureau faced a growing external political problem. Since its inception in 1912, the DeKalb County Farm Bureau, like most county farm bureaus, had a close and mutually beneficial relationship with the U.S. Department of Agriculture. The passage of the Smith-Lever Act in 1914 and the creation of the Cooperative Extension Service cemented that relationship to such a degree that the two organizations, one private the other public, became largely indistinguishable in providing DeKalb-area farmers and their children with services and educational programming. Such a close bond, however, occasionally raised problems. The first time it became an issue was when the Extension Service objected to advisor Tom Roberts dividing his time between the Farm Bureau and the DeKalb County Agricultural Association. Once that issue was resolved, the Extension Service and the DeKalb County Farm Bureau continued operating together as they had for years. Moreover, during the decades after the end of World War I, the American Farm Bureau Federation, the national umbrella organization of farm bureaus founded in 1919, evolved into the most powerful farm organization in the country. Because of the close connections between the publicly supported county extension agents and the private membership county farm bureaus, the American Farm Bureau Federation had a symbiotic relationship with the federal government that was unique among farm organizations.

By the middle of the 1940s, however, that close relationship came under increasingly hostile scrutiny in Washington, as well as by rival farm organizations such as the Farmers' Union and the Grange. As early as 1945, for example, some representatives in the Illinois state legislature wanted to separate the state extension service from private organizations, most obviously the Farm Bureau. The crux of all the critics' arguments was that the government's de facto endorsement of the Farm Bureau and the commingling of public bureaucracy with private interests were detrimental to the public interest because the federal government was using taxpayers' money to subsidize local farm bureaus. The charge was probably an accurate characterization of many farm bureaus in the nation, but one that did not reflect the situation in DeKalb County. On the contrary, the Extension Service in DeKalb County relied heavily on the DeKalb County Farm Bureau financing and other support for its programming. Uncertainty about a relationship between the Farm Bureau and the Extension Service and rumors about how the impasse would be finally resolved resulted in an uneasy operating climate for farm bureaus and frequent turnover of farm advisors. Uncertainty about the relationship and who was the supervising authority at the DeKalb County Farm Bureau, for example, meant that the Bureau lost four farm advisors between 1951 and 1954.[18]

The issue came to a head in late November 1954, when Secretary of Agriculture Ezra Taft Benson issued a memorandum that ordered the separation of the Extension Service from the Farm Bureau. Among other points, the order stated that no county extension agent could receive free office space or other compensation from any farm organization, advocate that any farm organization is better able than any other group to carry out the work of the U.S. Department of Agriculture, or solicit membership in any specialized organization of farmers. Although the language of the order was couched in general terms, it was clearly aimed at farm bureaus nationwide. The DeKalb County Farm Bureau, like most successful bureaus, was not happy with the "divorce." Former farm advisor Roberts expressed what was on the minds of many: "I don't like this divorce proceeding between Extension and Farm Bureau. I still think the Farm Advisor should be in charge of everything."[19]

The breakup also highlighted underlying tensions that existed within the Farm Bureau about its own identity. On the one hand, the close relationship by which the Farm Bureau provided money and facilities to the extension agent to carry out programming and services had worked well in DeKalb for more than forty years. The Bureau's financial success, due primarily to its close ties with the prospering DeKalb County Agricultural Association, enabled it to hire the very best advisors, carry out its traditional educational and service functions, and experiment with innovative new offerings, such as the cold-storage locker service. Those with longer memories, however, undoubtedly recalled that the original Soil Improvement Association had grappled with this same question of the appropriate relationship between their private organization and public agencies, opting for some private-public affiliation under the terms of the Smith-Lever Act only because the Soil Improvers desperately needed public assistance to make their fledgling organization succeed. What had begun as a marriage of convenience evolved into a close bond, but, in the post–World War II era, such relationships between government agency and private farm organization could no longer be sustained against the rising tide of public criticism.

In DeKalb County, Farm Bureau directors were determined to maintain their version of the farm advisor program but had no choice but to comply with Secretary Benson's order. To resolve the impasse, the Farm Bureau and the state extension service modified their agreement for extension support in such a way that the Bureau and the Extension Service could continue to provide all the services that Bureau members had come to expect. Under the new plan, rather than provide the extension advisor direct salary, office space, secretarial assistance, the use of a car, telephone, and other support, the Farm Bureau made contributions into a trust fund at the University of Illinois, which was, in turn, used to support extension services in DeKalb. The farm advisor also was no longer answerable to the Farm Bureau board, but rather to a state extension council appointed by the university. At the county level, a new DeKalb County Agricultural Extension Council and Home Economics Council, both comprised largely of Farm Bureau officers and members, offered advice to the farm and home advisors, assisted in selecting new advisors, evaluated their work, and helped with program planning. In the new Farm Bureau headquarters, the advisor's office was moved down the hall from its original and prominent place near the entrance and a partition constructed to further distinguish the advisor as simply another among many agricultural service tenants in the building. Patrons seeking services from the advisor were encouraged to enter the building through different doors that routed them directly to the advisor's office without passing the any of the Bureau's administrative offices. Despite these structural and cosmetic changes in their relationship, the close ties between the Farm Bureau and the farm advisor continued. As time went on, the Bureau further assisted the county's farm advisors with indirect support that skirted Secretary Benson's order, such as paying tuition and travel expenses to professional conferences, which, in turn, made DeKalb's farm advisors even more knowledgeable and successful. The transition of the farm advisor as an employee of the Farm Bureau to the new format of the County Agricultural Extension Council went so smoothly that few outside the Farm Bureau leadership even noticed anything but the most superficial changes.[20]

The "great divorce" did little to change the manner in which the DeKalb County Farm Bureau conducted its business, and during that readjustment period the Bureau continued to offer programs designed to help farmers succeed in a rapidly changing world dominated by new agricultural technologies, market realities, demographic changes, escalating social expectations, aggressive agribusinesses, and modern farm business techniques. New conditions in many parts of rural America, like the encroachment of large cities into farm counties and the growth of small towns into small cities, increasingly blurred historical distinctions between town and country communities. In the 1950s, while the DeKalb County Farm Bureau maintained its commitment to solving farm-specific problems, it also turned its attention outward in a much greater degree than ever before to issues that were important to all county residents, such as the development of a new junior college, improved hospital services and comprehensive health insurance, and expanding youth services to include urban children.

The person who most embodied this robust new Farm Bureau attitude was Howard Mullins, whom the Bureau membership elected as their ninth president in February 1956. Mullins was a successful farmer from Shabbona Township and had firsthand experience with the agricultural difficulties of the 1930s and the wartime stress that followed. Like most young agricultural men of the era, he had grown up with the 4-H. He had a forceful personality, which attracted the attention of Bureau leaders, and charm enough to woo even his most skeptical detractors. He was also a protégé of Henry Parke—Mullins was fond of quoting Parke's homespun wisdom, such as, "If you take enough steps in the right direction, you will eventually get there," "Have all the fun you can," and "Get the most out of your work."[21]

Until Mullins, Farm Bureau presidents generally worked to help the Bureau achieve its full potential within its traditional philosophical bounds of agricultural education and service. Mullins, like his predecessors, was equally interested in keeping the 3,300-member Farm Bureau the sort of grassroots organization its founders envisioned. "I sincerely want Farm Bureau," he wrote, "to be what you farmers want it to be. No other Farm Bureau in Illinois is doing as much for its members as your organization. . . . All our action will be by the democratic process and together we will back the majority." On another occasion he noted simply that the "Farm Bureau was an organization of farmers set up to help farmers and to be run by farmers." Yet, despite his populist-leaning rhetoric, Mullins also recognized that "we have reached a new era in Farm Bureau." Although he cautioned that "[w]e must not leave it to leaders in top positions or hired employees to formulate our policy," Mullins clearly represented a new generation of activist-minded Farm Bureau leadership and quickly emerged as the most dominant president in the history of the DeKalb County Farm Bureau and eventually became one of the most outspoken and controversial farm bureau chiefs in the nation.[22]

At the beginning of his tenure, however, Mullins appeared to be cut from similar fabric as his predecessors. One of the first programs that Mullins pursued as president was the Bureau's new Livestock Marketing Association, a marketing cooperative similar to the Bureau's livestock shipping associations of the 1920s. During the 1950s, national demand for meat declined as Americans changed their shopping and eating habits and foreign sales declined. Moreover, federal agricultural policies did not offer livestock operations the same price support and profit stabilization mechanisms that benefited crop farmers in the postwar period. Under the circumstances, DeKalb livestock growers found it increasingly difficult to remain profitable. It became evident to the Bureau's board of directors in 1956 and 1957 that a marketing association might help DeKalb stockmen reduce marketing costs and negotiate more favorable sales terms by pooling a large volume of stock and selling their output collectively. The Bureau's Livestock Marketing Association, which was independent from but affiliated with the Illinois Livestock Marketing Association, sold stock to Bureau members to construct a livestock shipping facility east of DeKalb to help area swine producers cut their costs, increase their profits by controlling a larger market share and selling in high volume to regional retailers, provide financial incentives to area producers to keep individual farm production high, and help offset the soaring fixed costs of production borne by every stockman. The association also arranged the sale of livestock to packers across the nation willing to pay good prices for large lots of DeKalb's high-quality swine. By 1960, the Livestock Marketing Association was successfully buying and selling swine, cattle, and sheep and providing area stockmen valuable tools to maximize their profits.[23]

The first signal that Mullins was taking the Bureau in a new direction was his approach to national agricultural policy. Although the DeKalb County Farm Bureau had encouraged its members to support McNary-Haugen proposals of the 1920s, state tax reform measures, public utility and insurance regulations, and a host of agriculture-friendly initiatives, historically the Bureau had been reluctant to endorse energetically contentious agricultural proposals at the national levels, fearing, perhaps, that political controversy might polarize members and threaten the organization's educational and practical offerings. Agricultural advisors, beginning with Eckhardt himself, had also distanced themselves from controversial policy debates that might undermine the Bureau and instead focused their attention, and their credibility as experts, on agricultural service. Mullins, however, adopted a wider vision of Farm Bureau responsibilities. For him, the Bureau could be more than a service agency for DeKalb County—it could be the collective voice of area farmers in important matters of national farm policy.

The change to agricultural activist came slowly, but he found a like-minded ally in the Bureau's farm advisor, E. E. "Al" Golden. Golden, like Mullins, saw the Farm Bureau as a vehicle for engaging area farmers in broad policy debates. Shortly after Mullins became president, for example, Golden initiated a series of programs and seminars for county residents that addressed several ongoing federal agricultural policy topics, including the Agricultural Adjustment Act, price supports and storage costs, conservation programs, and, most significantly, the new federal soil bank program proposed by Secretary Benson. In historical Soil Improvement Association style, these educational meetings were usually question-and-answer forums about national policies. The meetings were so successful that they quickly evolved into regular "policy development meetings," held regularly in towns conveniently located throughout county, at which Mullins, Golden, and Farm Bureau leadership could gauge members' viewpoints on a wide range of agricultural policy issues. Participation in the political arena and support of the Bureau's collective agricultural voice was elevated to a moral responsibility. "Only by participating in this effort to gain the best thinking of all Farm Bureau members," said one policy development meeting announcement, "can farmers fulfill their responsibility to themselves and their neighbors. It is imperative that Farm Bureau members face problems together rather than sit back and wait to see what the other folks do."[24]

The Farm Bureau's new trend toward activism began early in Mullins's tenure in office, when the Bureau backed the nation's soil bank program. The soil bank was a controversial federal program designed to supplement a new flexible agricultural price support program, first approved by Congress in 1954, which had replaced the fixed price-support programs that had been in place since the war. Fixed supports, which had worked well in Depression and war to keep farmers afloat, were an anachronism by the early 1950s that too often resulted in overproduction of supported crops and underproduction of unsupported ones. To many economic reformers in Washington, flexible supports seemed to be the best method of weaning farmers from price supports by encouraging them to change their economic behaviors, or by forcing inefficient and marginal farmers out of the business altogether. The flexible support program financially ruined thousands of farmers across the nation, and to soften the blow Secretary Benson proposed the soil bank program in 1956. Under the soil bank, which lasted until 1958, farmers leased some of their farmland to the federal government, which, in turn, put the land into grass or other soil-conserving but unmarketable crops. The bank did not remove the economic threats posed by flexible supports, but it extended the time that farmers had to convert to the new policy and thereby helped thousands of hard-pressed farmers remain on their land. In February 1957, after discussing the program with Farm Bureau members for much of the preceding year, Mullins and the Farm Bureau board of directors agreed to publicly support the program, noting that it had overwhelming support from the DeKalb County Farm Bureau members. They were not content, however, as they once might have been, to simply publish the position of DeKalb's farming community and encourage its members to send letters of support to Washington. The board of directors sent the Bureau's resolution of support for the soil bank to other county farm bureaus in the Midwest, the Illinois Agricultural Association, and the American Farm Bureau Federation, urging them to likewise support the soil bank plan.[25]

Although the soil bank ultimately had little impact on the largely successful DeKalb County farmers, the event signaled an important change in the DeKalb County Farm Bureau. Mullins, it was clear, was revealing himself to be an aggressive leader who was directing the Farm Bureau toward an activist role in shaping state and national farm policy. The core of his activism was his belief that existing agricultural organizations were ineffective in addressing the challenges confronting the agricultural sector in the postwar era, especially the serious economic problems associated with the huge agricultural surpluses that had built over years in part because of government price supports. Farm prices in the late 1950s were "dangerously low." "After allowing a reasonable return on their investment," noted former Secretary of Agriculture Claude Wickard, "they [corn belt farmers] do not receive a dime for their labor."[26] Economic conditions down on the farm, Mullins believed, demanded immediate action. Mullins and the Bureau's board of directors saw the need for new state and federal policies to keep farmers afloat. With the blessing of the DeKalb County Farm Bureau board of directors, members, and advisor Golden, and with an informal Bureau policy advisory committee offering support, Mullins was steering the Bureau, already one of the best-known and most successful county farm bureaus in the nation, to the forefront of agricultural policy making.[27]

Kishwaukee Service Company salesman William Stahl at the Pleasant Street bulk plant, DeKalb, Illinois, 1958. The other bulk storage facilities were located in Genoa and Waterman, Illinois.

A home economics 4-H club field trip to one of the DeKalb County Locker Service facilities, 1958.

A DeKalb County Locker Service deliveryman and his truck by the Service's home office at the corner of Fifth and Pine streets, DeKalb, 1960.

above—DeKalb County Exports Company elevator and barge loading facility on the Illinois River, near Ottawa, Illinois, 1978.

left—Unloading a grain truck at the DeKalb County Exports Company elevator, 1978.

Loading barges with "identity-preserved" grain at the DeKalb County Exports Company elevator, 1978.

The meatpacking facility and store for the Bureau's Country Pride Meats, located at 108 Harvestore Drive, DeKalb, mid-1980s. The packing plant continued the Bureau's history of involvement in processing and marketing locally grown meat and was a state-of-the-art facility when it opened in November 1980.

A copy of the *Farmer's Point of View*, printed using soy-based ink, March 1991. Eckhardt was an early advocate of soybeans as part of a comprehensive crop-rotation program and published an article on the subject in March 1912. Eckhardt planted the first soybeans in the county only a few days after he began work, in June 1912. The next year, a few farmers followed Eckhardt's advice and planted soybeans, which they harvested as "bean hay" for livestock. For decades, the Bureau worked to promote soy-based products. Soyoil paints were introduced in Illinois in 1930 and were heavily promoted by the Kishwaukee Service Company and the Farm Supply Company for

Corn car, 1991. The Bureau is active in promoting ethanol, derived from corn, as a fuel additive.

The Bureau sponsors six "Associations" (DeKalb-Kane Cattlemen's Association, DeKalb County Corn Growers Association, DeKalb County Soybean Association, DeKalb County Lamb and Wool Producers, DeKalb Area Pork Producers Association, and Kishwaukee Dairy Herd Improvement Association) that work to educate farmers and the public about their products and promote their industries. As part of their programs, the corn growers and soybean associations plant comparison plots, while the Cattlemen's Association works to win over the public in more direct ways.

below—The second headquarters of the DeKalb County Soil Improvement Association, 320 North Fifth Street, DeKalb, Illinois, 1915. Beginning in 1918, the "old North School" building also housed the DeKalb County agricultural association, a seed warehouse, and a seed-testing laboratory.

above—The third home of the DeKalb County Farm Bureau, at the corner of Sixth and Oak streets, DeKalb, dedicated August 29, 1951. The Bureau began saving money for the building in the late 1930s, but construction was delayed by World War II. The Bureau broke ground for the building May 5, 1950. Upon completion, the building was home to the Farm Bureau, DeKalb County Grain Company, DeKalb County Home Bureau, DeKalb County Locker Service, Fox Valley–Joliet Production Credit Association, Illinois Agricultural Auditing Association, Illinois Agricultural Association Insurance Service, Kishwaukee Service Company, National Farm Loan Association of DeKalb, Northern Illinois Breeding Co-Op, Production and Marketing Administration, and the Soil Conservation Service. It was truly a one-stop agricultural service center.

Architect's elevation of the Bureau's fourth home, located at the corner of Bethany and Peace roads in Sycamore, Illinois. The Center for Agriculture was dedicated September 12, 1996.

In the spring of 1960, Mullins, together with Paul Montavon, an influential DeKalb County dairyman, and Purdue University agricultural economists J. Carroll Bottum and John Dunbar, drafted a short summary of what they believed were the most important agricultural policy initiatives required to solve "the critical income and financial situation of Midwest farmers." While it quickly became a sort of DeKalb County Farm Bureau policy manifesto, it was hardly a revolutionary proposal. "We recognize that any practical solution to the farm problem must be based not only on the needs of farmers but also on the interests of society in general." Although increased exports and expanded use of agricultural products at home would help ease the crisis, new markets alone could not solve the farm problem. The authors stressed "conserving the soil and water, assuring abundant food supplies, upgrading the diets and protecting our wildlife, forestry and future use of the land." To achieve those ends, they envisioned a program that would retire from cultivation about 80 million acres, roughly 17 percent of the nation's total farmland. If such a program was mandatory, they argued, it would need to be enforced so that the retired land would not be shifted to other lines of production. If voluntary, federal payments needed to be high enough to offer incentive for land retirement, and likewise strictly enforced. Finally, the program needed to be put into effect immediately. In short, while traditional federal farm subsidy programs purchased farm products after they were grown, resulting in large farm surpluses and depressed prices, Mullins and his colleagues sought to prevent the crops from being grown in the first place and thereby reduce farm supplies and boost prices. "We feel that past government programs have been costly and that they have not solved the farm problem. We feel that our proposal is the major step to be taken at this time and that, properly administered, it can key agricultural production nationally to anticipated use, and thereby raise farmers' prices and income by a program of producing for a market. We are confident that the ultimate cost of this program will be far less than the cost of present or past programs."[28]

In May 1960, Mullins and Montavon organized a meeting of sixty-seven agricultural leaders from seven midwestern states to meet at the DeKalb County Farm Bureau's new headquarters and discuss their proposed outline of action. "We are here," said Montavon in the opening address of the meeting, "to pool our thinking on the major problem that faces farmers as a group. . . . We are here to exchange ideas. That means that each of you must have an opportunity . . . to express his own opinion. We didn't invite anyone here to sell something. We invited you here to exchange ideas." Mullins and Montavon were also keenly aware that they needed to stay above partisan politics and reach across party lines as a nonpolitical and unaffiliated organization—the May 26 meeting was attended by Republicans, Democrats, and representatives of the three leading farm organizations. "If [the farm problem] could have been solved by a farm organization or by a political party, I think it would have been solved long ago. . . . Let's be honest about it. We have two main political parties. One party comes up with a plan. Do you honestly think the people of the other party are going to [agree] or do you think the people of the other party are open-minded and admit that it is good[?] It has gotten to that point with different farm organizations, each with a different approach. And you wonder why Congress is still working on it and never accomplishing much." And, with a touch of humor, Montovan said simply, "If I didn't have a good answer, I would be home planting corn." Regional newspapers, agricultural journals, and area radio networks covered the May 26 meeting. The representatives at the meeting styled themselves the United States Farm Policy Council and opened membership to anyone, including "business, professional or other groups whose economic fortunes are closely tied to those of agriculture." A specific invitation was extended to "staff members of land grant colleges who have taken interest in matters of this general nature." They also endorsed the acreage reduction program outlined by Mullins and Montavon and elected an executive committee of eleven members, with Mullins as chair, to pursue their new program. The mailing address for the Farm Policy Council was the DeKalb County Farm Bureau Building.[29]

The U.S. Farm Policy Council's *Statement of Functions and Objectives*, the organization's "constitution," set out its underlying philosophy in its preamble. "At various times," it stated, "[the farmer] has suffered economically as well as through decline of the esteem in which he is held by the general public. The typical farm producer has no desire to be a 'ward of the government.' On the contrary he is an individualist with an inborn aversion to acceptance of anything that smacks of subsidy or special treatment." Few, however, could disagree that it was important to solve the economic crisis in the heartland. The key to unraveling the crisis was education. "[A] basic essential to such a solution is understanding the true facts by everyone including farmers themselves, urban people, lawmakers, and the public at large. Most farmers when properly informed will be willing to subordinate individual considerations of self-interest to the national group interest. . . . There is need for a mechanism to inform people everywhere, urban and rural alike, of the facts pertaining to the economics of agriculture and to its important relationship with other segments of our economy." The council also stressed the national security benefits from such an education program. "[Farmers] will accept the broad view that our national strength and security are dependent upon cooperative effort toward fairness, equity and justice to the members of all our economic groups."[30] Finally, the council emphasized that cooperative action was the best means of attaining their goals. "There is . . . need for a systemic procedure designed to promote unity of effort among farmers and organized farm groups in the attainment of their common objectives." Although the council was breaking new ground in its appeal to farmers to work knowledgeably and cooperatively to support its crop reduction proposal, its methods were drawn straight from the history of the Soil Improvement Association and the farm bureau movement—education was the key to empowering farmers, and cooperative action could bring about change.[31]

On July 19, 1960, the nonpartisan and unaffiliated Farm Policy Council held its next major meeting, again in DeKalb, to develop a plan of action to raise public awareness of the agricultural sector's stagnant economy and persuade Congress to adopt their program. It was an uphill battle. The American public took the agricultural sector for granted and showed little concern for struggling farmers, whom city dwellers assumed were doing well. There were few sympathetic editorials or articles in the media. To raise public awareness, the Farm Policy Council proposed a public relations campaign designed to educate the nation about the farm problem and the council's proposed solution through a comprehensive public education initiative. "A well organized program participated in by all agencies, organized groups, educational units, and individuals with direct or indirect interest in agriculture, would have great beneficial effect in informing people of the success story of agriculture, in shaping their attitudes toward dealing with the economic problems of farmers, and in developing programs of action which would benefit not only farmers but the economy of the entire nation." The more than 1,000 midwestern farmers who attended the July meeting embraced the council's educational ideal and were overwhelmingly in favor of a mandatory acreage reduction plan. Mullins announced that the council's executive committee planned to meet with Republicans that month at their convention in Chicago and had already presented the council's general views on acreage reduction to the Democratic National Committee at its convention in Los Angeles a month earlier. Other regional meetings followed in Omaha, Nebraska, in October, Des Moines, Iowa, in December, and Holdrege, Nebraska, in March, 1961.[32]

The Farm Policy Council's dual emphasis on public education and political pressure was an aggressive strategy that ran contrary to the national farm bureau position. In December 1959, a few months before the Farm Policy Council began its campaign for an acreage-reduction policy, the American Farm Bureau Federation, the national umbrella organization for farm bureaus across the nation, had rejected a similar proposal to pursue acreage reduction as a solution to the troubled farm economy. Following the lead of the national organization, the state farm bureau organization, the Illinois Agricultural Association, also dropped its support for the acreage-reduction issue. Unlike his predecessors in the DeKalb County Farm Bureau, who seldom criticized other agricultural organizations, Mullins was openly hostile to the American Farm Bureau Federation and state organization, arguing that they had done too little to develop and support policies that would help the common farmer. "The Farm Bureau on the national level," Mullins once commented, "was doing nothing." He was unimpressed, and often said as much publicly, by the Illinois Agricultural Association representatives, who defended what Mullins believed were outmoded philosophies and ineffective policy responses—it is not surprising that such blunt and critical activism made Mullins unpopular at Illinois Agricultural Association meetings. Charles Shuman, president of the American Farm Bureau Federation, led a campaign to kill the acreage policy proposal within the American Farm Bureau Federation and remove it from the table as a policy alternative. He also used his Washington influence to attack the Farm Policy Council.[33]

Mullins and the Farm Policy Council leadership were undeterred. With support from the DeKalb County Farm Bureau, the council went on the offensive, working to generate grassroots support for its initiatives throughout the Midwest. Mullins met with politicians in Washington to learn what was possible politically and consulted with economists to ensure that their ideas were sound. Moreover, the council's efforts at raising awareness for their plan were succeeding. In its August 1961 newsletter, the Farm Policy Council highlighted samples of correspondence that reflected the reach of its message. "Your organization has done a good turn for the agricultural industry," wrote a member of the Washington state National Association of Wheat Growers, "and I hope that you can keep up the good work. Our groups think very closely together and I am sure that the fine cooperation can be continued." A U.S. Department of Agriculture official wrote, "I have read your letter with great interest. The brief history of the Council is very informative, and I wish to say that this Council should be a constructive force in formulating a sound farm policy and program." An Arkansas cotton farmer wrote, "I have been following your Council's activity with great interest, and realize that the people with whom you are associated in this program have the interest of heart of the farmers in this country." A midwestern banker noted, "You are to be commended for keeping this matter before farmers and farm leaders. We think it is extremely important that views be crystallized by meetings of this type so that plans can be formulated for a useful and practical farm program." Even a doctor from Brooklyn, New York, wrote, "I do hope many public spirited persons will participate in the Council's program and make it a real force. . . . I should like to help by being a dues-paying member, and I shall lend support where I can."[34]

Mullins especially seemed to thrive on the controversy and the rough-and-tumble arena of shirtsleeve agricultural politics. At the same time, however, he was not always the most diplomatic spokesman for his cause. Although he was clearly able to capture the imaginations of rural folk, his blunt, plainspoken style did not play as successfully among the national and state farm bureau leadership. He and Shuman were barely on speaking terms and did little to conceal their mutual dislike. Shuman did all he could, directly and indirectly, to derail the council's initiative and undermine Mullins's credibility. At the DeKalb County Farm Bureau annual meeting in the winter of 1961, with the Illinois Agricultural Association president in attendance, a clearly frustrated Mullins at one point openly threatened to simply buy the Illinois Agricultural Association. The real sting in the threat, however, was its obvious reminder that the DeKalb County Farm Bureau was probably more powerful, financially and politically, than the state organization. Though such heated public comments were rare, they hardly won Mullins, the DeKalb County Farm Bureau, or the Farm Policy Council many friends at the state level.[35] In the end, the Farm Policy Council needed neither national nor state support. It was a grassroots policy-making campaign that gathered momentum from farmers themselves, rather than from the established farm organizations like the American Farm Bureau Federation. Although the leadership of the American Farm Bureau Federation remained skeptical of the upstart Farm Policy Council and its vocal DeKalb spokesman, it was clear that rank-and-file Bureau members were excited about the possibility that their organization was so directly shaping new federal action.[36]

By early 1961, the Farm Policy Council directed its attention away from organizing and educating the public generally about the farm problem and toward a proposed feed-grain program, introduced in Congress by President Kennedy on February 16, 1961, as the most direct way of achieving its goals.[37] The object of the legislation was to reduce surplus feed grain and thereby raise farm incomes. In the short run, that meant reducing government surpluses at rates that did not undercut farm incomes. In the long run, the council advocated that the federal feed-grain program be solidified by a land-retirement program designed to hold production in line with consumption. Council leaders were anxious to note that this bill would not be a "government program." Rather, "it is our program as farmers—the government's part is merely to set up the machinery through which we may operate it. . . . [We should] regard it as *our* program, get behind it, make it work. By doing so we shall gain experience which will enable us to work out an improved permanent program for 1962 and the years that follow."[38]

The council's leaders were so confident of its success that they began formulating long-term policy goals for their organization. In addition to acreage reductions, they extended their vision to include comprehensive land-use and water-resource planning, support of the "family farm," fundamental tax reform, improved opportunities for farm youth, a national food cache program for the poor, better international marketing, a "food for peace" program, an improved Cooperative Extension Service, integration of agricultural perspective into nonfarm national policy debates, and self-empowerment of farmers to design and initiate broad national policy reforms. The insurgency also attracted considerable attention in Washington. In May 22, 1961, Mullins testified before the Senate Agricultural Committee about the economic problems of American farmers and the acreage reduction idea. His comments caught the attention of powerful Kennedy administration insiders, such as Secretary of Agriculture Orville Freeman, Freeman's economic advisor Willard Cochrane, and Undersecretary of Agriculture John Schnittker.[39] The following month, on June 21, 1961, President Kennedy appointed Mullins to one of the Republican seats on the National Agricultural Advisory Commission. It was a surprise to everyone, especially Mullins, the Shabbona farmer who suddenly found himself at the very center of national farm policy making and consulting with the Kennedy administration. It is likely that Kennedy chose Mullins not only because of their mutual support for the feed-grain program and the acreage reduction concept, but also because Mullins was an agricultural Republican clearly at odds with the powerful and politically entrenched Republicans in the American Farm Bureau Federation. Throughout 1961, the administration's Agricultural Advisory Commission developed a series of policy proposals to address the economic conditions in the agricultural sector. As a commissioner, Mullins was instrumental in persuading the Kennedy administration to steer away from a mandatory feed-grain reduction program, which was overwhelmingly unpopular with farmers, and to adopt a voluntary program that included many of the core goals of the Farm Policy Council.[40] Mullins, already president of one of the most successful and powerful county farm bureaus in the nation, had risen quickly to become one of the country's leading agricultural spokesmen.[41]

Mullins's work in national agricultural politics was a unique role for DeKalb County Farm Bureau presidents. Although Eckhardt was involved in state affairs as the emergency wartime state seed corn administrator, and later as director of grain marketing for the Illinois Agricultural Association, few Farm Bureau advisors, officers, or directors had spent much time in the political limelight, and none had risen to such national prominence as Mullins. Yet, for all the emphasis on national policy making, Mullins and the Farm Bureau also took time in 1962 to reflect on five decades of Bureau leadership in the county. At the fiftieth annual meeting of the Farm Bureau, held at the Egyptian Theater in DeKalb in February 1962, Mullins announced that the Bureau was planning a birthday party, to be held on June 14. The celebration, "Progress Through 50 Years of Farm Bureau," was the largest in DeKalb since the first Farm Bureau celebration in 1922. More than 20,000 people lined the sidewalks of DeKalb to watch the parade of 130 floats portraying changes in farming since 1912. Antique tractors, trucks, and cars carried local dignitaries and former Bureau officers. Floats by local clubs included such rural-theme historical titles as "Down on the 1912 Farm," "Rural Electrification," "Modern Living for the Hired Man," and "Men of the Farm Bureau, A Golden Age, a Grandfather." The Farm Bureau honored the handful of charter members of the Soil Improvement Association who were still alive. The DeKalb County Swine Association hosted a barbecue for thousands, while the DeKalb Municipal Band played old-time favorites for the crowd. The highlight of the evening, just as it was during the celebration forty years earlier, was a pageant that celebrated the history of farming in northern Illinois in fourteen acts. "All comments received," reported the *DeKalb Daily Chronicle*, "indicate that this is the greatest achievement in agriculture in DeKalb county. . . . The 50th anniversary celebration represents a tremendous achievement that probably can never be repeated with the same historic individuals present."[42]

Throughout the 1960s, the Farm Bureau continued to provide a broad range of educational, social, and service programming for its members and the community. Mullins, however, was concerned about the new direction that the DeKalb County Farm Bureau was moving—it is ironic that Mullins articulated this fear when he was, after all, one of the most influential leaders moving the Bureau away from its traditional role. He lamented that the Bureau was straying too far from its core educational mission of making DeKalb farmers more financially secure and put agriculture on an equal footing with the other sectors of the nation's economy. Mullins also feared that the Bureau was operating too much like a business, selling insurance and other farm products, and losing contact with the Extension Service and state universities. Politicians in Springfield and Washington seemed to be less concerned about supporting agricultural issues as the percentage of farmers in American society declined. "We are inclined," said Mullins, "to let small groups of people or even individuals do most of our thinking for us."[43]

To remedy this trend, the Bureau looked to its traditional educational and service missions and initiated new programs to address contemporary issues. One of the Bureau's most timely new educational programs focused on farm safety. Farming was always a dangerous profession. During the mid-1960s, annually 8,000 farmers died and tens of thousands more were disabled—it was the third-most-dangerous occupation in America. Injuries caused by complex and dangerous machinery, traffic accidents, and home accidents were common. Moreover, Rachel Carson, Ralph Nader, and a host of other popular public health and safety advocates called for the nation to reconsider the safety of the potent new chemicals and insecticides that were commonly used in agriculture. Americans began to understand that many of the chemicals were more dangerous than originally believed and posed significant threats to the long-term health of agricultural workers. In 1962, the Kennedy administration declared the week of July 22, 1962, to be National Farm Safety Week. The event, titled "Farm Safety—At Work and Play," was cosponsored by the National Safety Council and the Department of Agriculture. The DeKalb County Farm Bureau had been concerned with safety for years. As early as the 1920s, for example, the Bureau encouraged farmers to buy insurance to help offset personal casualty losses; the Home Bureau presented units on home safety; and the *DeKalb County Farmer* published occasional pieces on farm and highway safety. Historically, however, the Bureau did not undertake regular farm safety programming, probably because most rural folk simply accepted the risks of farming.

During the 1960s, however, the new American concern with public and personal safety prompted the Bureau to create a regular safety committee, whose motto was "Keep Your Guard Up—Stop Accidents," and to sponsor numerous workshops on farm safety. Mullins, himself a victim of a farming accident with a hay-baler that disabled his right hand, was an enthusiastic supporter of the new program. He believed that most accidents were the result of carelessness and haste and that such an education program would go far to raise safety awareness and reduce farm accidents. Lifting his mangled hand years later, Mullins said simply, "I had good reason to promote farm safety."[44]

Although Mullins and others sometimes lamented changes in the Bureau, the leadership also understood that their organization needed to continue its financial success if it wanted to sustain educational programming. The directors made a variety of important business decisions during the decade to shield the Bureau's resources. In 1961, the Locker Service suffered a $14,000 loss for the year, signaling the beginning of the end for what had once been an important Farm Bureau service. The Bureau finally liquidated the service in April 1965, a victim of home freezers and the trend toward convenience supermarkets. In 1966, the Bureau's insurance affiliate, Country Companies, began to offer members a long-term capital growth mutual fund promoted by the Illinois Agricultural Association. The idea was based on the premise that farmland, which had been the traditional investment avenue for most farmers, was expensive and difficult to buy and that farmers needed alternative and flexible investment opportunities. The Bureau's affiliated Kishwaukee Service Company, which had been delivering petroleum products in DeKalb County since the late 1920s, merged with its counterpart in Kane County to form Northern FS, Inc., in late 1969. The new offices were located in the Farm Bureau building.[45]

The Bureau also made many small changes. In 1963, the Farm Bureau raised dues, from $21 to $23 annually. The resulting several thousand dollars in yearly revenue helped offset Bureau costs, but it is a testament to the success of the Bureau's early acquisition of DeKalb County Agricultural Association stock and financial management over the decades that it could provide so much programming and ask for so little in dues. The Bureau improved its telephone system so that the extension agent could provide specialist advice without incurring travel time and expense. The Bureau regularly supported road construction and improvements in the county, and better roads meant easier access to town. Old rural habits faded. Farmers who once came to town to shop only on

Saturdays now found it more convenient to take care of their affairs during weekday business hours. In December 1964, the Farm Bureau closed on Saturday and Sunday for the first time in its history. In 1965, the Bureau offered its members bail bond certificates. In 1966, the Bureau purchased the north side of the block next to its building from the DeKalb County Agricultural Association for $50,000 to expand its parking lot. Also in July 1966, the Bureau sponsored its first Farm Bureau Play Day, a social event that featured a picnic, horseshoes, trap shooting, and a golf tournament. The annual Play Day quickly became a high point in the summer social calendar for many Bureau families. Throughout the decade, the Bureau helped arrange training for Peace Corps volunteers, International Farm Youth representatives, and other visitors who studied modern and historical farming practices that they could use to help farmers outside the United States.[46]

One of the most important educational initiatives during the decade was the Bureau's support for a junior college in the county. Junior colleges were especially attractive to farming communities. For generations, farmers' institutes, Soil Improvement Association and Farm Bureau seminars and workshops, practical demonstration work, 4-H, and other formal and informal learning opportunities provided farmers with agricultural training. Formal education beyond public high school, however, was a luxury in terms of both money and time that few farm families could afford. As farming became more complex during the twentieth century, however, it was increasingly necessary for farmers to master a host of new topics, such as economics and business management, if they hoped to succeed. A junior college was an alternative to higher-priced four-year colleges and universities, offered advanced training in practical subjects at flexible times, and welcomed part-time students. Junior colleges also offered education beyond high school for rural youth who wanted nonfarm jobs or older students changing careers. It had the added advantage that it would be located in a rural area so students could live at home and continue to help with the family farm. Beginning in 1965, when a junior college was suggested for DeKalb County, the Farm Bureau was an ardent supporter of the plan. It informed area voters about the advantages of the college and pledged thousands of dollars toward the founding of Kishwaukee College. With strong support from parts of the county, rural voters passed a referendum on January 14, 1967, and elected a seven-member board of directors the following month to supervise the college's formation. The majority of members of the original board of directors were either farmers or men who had deep ties to the farming community. Local families, for the first time, had access to a postsecondary education that was oriented toward providing practical training to meet the needs of modern rural America.[47]

The Bureau's new policy development committee was also busy during the decade. Changing national demographics meant that fewer American farmers were producing more than ever before. Rural folk had less voice on public issues than during the nation's past, when a majority, or even a significant minority, of the citizenry worked the land. Local, state, and national farm bureau organizations stepped in to articulate and amplify farmers' concerns about agricultural policy. Such political action mirrored the Bureau's collective spirit in other areas, such as marketing and purchasing. Regular Farm Policy Committee meetings kept farmers informed about important agricultural issues and closely monitored member concerns. The DeKalb County Farm Bureau's policy committee spoke for the membership on local measures, such as the Bureau's support of a prohibition against dumping garbage on private land. It also articulated positions on state and federal measures that governed farms including land, product, fuel, and equipment taxes; the exemption of farmer haulers from Interstate Commerce Commission regulations; exclusion of farmers from compulsory worker compensation statutes; the federal Agricultural Adjustment Act and its regulations; and the revision of Illinois's antiquated state constitution.[48]

Farm bureau political activism, especially by the American Farm Bureau Federation at the national level, led to increasingly hostile public critique of the Farm Bureau. In 1967, Representative Joseph Resnick (D-NY) launched a congressional inquiry into rural poverty that led quickly to an investigation of the American Farm Bureau Federation and revealed many troubling practices. The American Farm Bureau Federation often strayed outside the agricultural policy arena and endorsed controversial social positions that further alienated it, and by association local farm bureaus, from the public. The American Farm Bureau Federation, for example, opposed the Voting Rights Act of 1965, generally opposed minimum wage laws, called for the elimination of the Department of Education, opposed the Equal Rights Amendment, and attacked labor organizations, especially rural labor groups such as the United Farm Workers. Moreover, the American Farm Bureau Federation's historically anticommunist rhetoric and general hostility to the sweeping social changes occurring during the 1960s kept the federation firmly entrenched as a voice of the political right. To many Americans, such activities made the American Farm Bureau Federation—and, by association, farm bureaus everywhere, including DeKalb—seem out of step with the times and opened it to criticism that it was simply the political voice of reactionary farmers.[49]

The move toward greater political activism, however, could not change the economic realities that faced DeKalb farmers during the 1960s. Farmers in DeKalb County were highly productive; in 1968, for example, DeKalb County had an average corn yield of 105 bushels per acre, which was the highest in the state. Such large yields, however, contributed to the surplus that depressed farm incomes. Mullins and other Bureau leaders and members understood that, although the federal government remained committed to its farm-assistance programs, it was not likely to boost farm support levels sufficiently to offset flagging incomes. Political and popular interest in the farming sector had declined in the postwar decades. In the reform era of the 1960s, a constellation of social and political issues competed with farm troubles for public sympathy. At a Farm Bureau policy discussion meeting held in DeKalb on September 24, 1968, one of the most important such meetings of the decade, representatives of regional farm organizations, commodity-marketing groups and agribusiness leaders discussed two basic questions: "Why is agriculture in trouble?" and "What can we do about it?" They were both relevant questions. At the time, even DeKalb County farmers were finding it difficult to make much profit on their enormous corn crops, and many were losing money. It seemed to make little sense that surpluses should be so high when, according to estimates at the time, half the world's population did not have enough food. For the assembled farm organization representatives, there were few alternatives: reduce production through land retirement, as the Farm Policy Council advocated, or find new markets to absorb the surpluses.[50]

In response to the stagnant economic conditions, the Farm Bureau encouraged its members to be politically astute and work within the existing federal agricultural programs for an orderly adjustment of agricultural production to meet market need at competitive price levels. The Bureau also focused more attention on ways of increasing market share for its members. Developing new markets was a natural response. Since the 1920s, the DeKalb County Farm Bureau had helped local farmers increase their sales through collective action. Cooperative actions—such as the first seed purchasing program of the pre–World War I period, shipping associations of the 1920s, the DeKalb County Grain Company, and the Livestock Association during the late 1950s—were focused on increasing farm productivity, efficiency, and regional market share for

DeKalb County agricultural output. During the 1960s, the Bureau continued its traditional programs and even improved offerings in such areas as insurance, youth outreach, financial services, soil testing, and soybean cultivation. Mullins and his allies in the Farm Bureau, however, had higher aspirations for their organization. Cooperative marketing had worked reasonably well in the past, but harnessing local, or even regional, demand was clearly no longer enough to offset losses resulting from oversupply. Foreign markets, Mullins and the Bureau leadership believed, held the key to long-term prosperity of area farms.

The idea of expanding DeKalb County's international market profile reflected larger economic trends during an unusual time in the history of American agriculture. Although high farm production was setting records and driving down prices, the national and international conditions seemed to promise greater consumption and a greener financial future for farmers. At home, American politicians and Great Society humanitarians during the 1960s redefined poverty and "discovered" a new population of hungry people living in the United States. These domestic poor were potential consumers for surplus farm products purchased and distributed by new governmental welfare programs, such as food stamps and school breakfasts and lunches. Feeding hungry people could yield good financial returns for farmers. Poverty was rampant outside the United States, too. During the 1960s, there was a tragic shortage of agricultural products, especially in developing nations. Demographers predicted that the world population would outpace food production, especially in underdeveloped nations and behind the Iron Curtain. A relentless 2 percent annual increase in world population meant that the world's population would double by the turn of the twenty-first century. American farmers saw their work as a humanitarian necessity, as well as a livelihood. This was hardly a new phenomenon—farm production, after all, had been vitally important during two world wars. What made this world food crisis different was that it occurred during a time when the nation was fighting a war in Vietnam without mobilizing American society directly, the rest of the world was generally at peace, and the domestic cold war economy appeared large and stable enough to extend a hand to even more of the world's destitute. Finally, even prosperous European nations, many of which relied on antiquated farming and marketing infrastructures, were becoming more dependant on American farm imports to sustain their expanding industrial economies. Still other foreign buyers were interested in importing high-quality American farm produce for special purposes, such as food manufacturing and distilling, and found that American grains were less expensive, even with transportation costs and duties, than locally grown European products of similar quality.[51]

American farmers who saw domestic and foreign demand for their products accelerate eagerly supported the exploitation of international markets in particular and maximized domestic production to meet the rising new demand. By the middle of the 1960s, nearly 17 percent of total U.S. farm acreage production was exported. Forecasts suggested that annual cereal consumption would rise about 2 percent annually, from just over a billion tons in the late 1960s to 1.9 billion tons by the early 1990s. They predicted demand would outpace production by nearly 100 million tons a year. Anticipating a new and more prosperous business climate, farmers planted "fencerow to fencerow," bought more land, invested in new machinery, applied more chemicals and experimented with new ones, raised new strains of hybrid crops and livestock, and upgraded buildings and homes—expansion, after all, had always been a sure route to prosperity when the farming business cycle was rising. At the close of the decade, exports made the agricultural future look brighter than it had a few years earlier. Farmers' slowly increasing net worth improved farm morale, and many rural families found that were able to enjoy a similar living standard as townsfolk. It looked as if farming for the international market would be a growth industry during the 1970s, and the DeKalb County Farm Bureau wanted area farmers to be at the leading edge of the new trend.[52]

This improving economic horizon had important consequences for American farmers. New profitability and opportunity in the countryside slowed the exodus of farm children. Farm-related businesses prospered; however, it also caused farmers to change their perception of financial success. The distinction between wealth and the illusion of wealth became blurred. One central feature of the new farm economics was the dramatic increase in land values. During the 1970s, it was not uncommon for good farmland to increase 300, 400, and even 500 percent. Farm incomes, on the other hand, did not keep pace with the rate of land inflation. According to one economic calculation, many farms were producing less than 3 percent of their market value—in other words, substantially less than what a similar investment would earn at a bank or in other nonfarm investments. Consequently, only about a quarter of farm wealth came from farm income, with the rest reflected in the higher land values. Farmers began to see capital gains as an equally important, if not more important, part of their total financial status. These

paper values fueled an economic cycle rooted in inflation and high expectations. Equity financing became more common as farmers expanded their operations to meet the new demand and extract cash from
the equity of their valuable land. Lenders saw the soaring land values
and were eager to lend money. Farmers saw the tax advantages of debt
and comparatively low interest rates and rushed to borrow, fearing that
if they waited too long they would lose out on the bonanza. Such
thinking reflected faith that agriculture could sustain high, even exor-
bitant, land prices and that the boom times based on increased domestic and foreign consumption would finally make farming lucrative.
With total farm debt growing from under $50 billion in 1970 to $190
billion in 1984, marketing produce for income revenue became ever
more important to farmers who had to service their swelling debt to
sustain prosperity.[53]

The DeKalb County Farm Bureau's movement toward foreign marketing was not an immediate decision and had been slowly evolving since
at least the early 1960s. In 1963, the U.S. Department of Agriculture
asked Mullins to participate in the European-American Symposium on
Food and Agricultural Trade, in Rotterdam, the Netherlands. The meeting helped Mullins refine his knowledge of international grain demands,
international balance-of-payments issues, and the grain-exporting business. While in Europe, Mullins was disturbed by the inferior quality of
grain that other nations imported from the U.S. in comparison with the
grains produced by DeKalb County farmers. Much of the American grain
being sent abroad was damaged or contaminated with inedible matter
and was in general of a lower grade than what was commonly raised in
DeKalb. "I knew corn I produce and what my neighbors grow was much
better than what I saw," wrote Mullins. "I believed if this country didn't
do something quickly, it would eventually lose its grain market to other
countries." The following year, 1964, Farm Bureau leaders asked Golden
to participate in a twenty-one-day trade mission to eight European nations, led by the Illinois Agricultural Association. The group toured the
region and showcased Illinois farm products. Golden, too, discovered
what he believed were opportunities for American farmers to market
their production abroad. "Population growth [in Europe]," he reported,
"impressed me the most. The people must be fed and cannot produce all
they need: with our ability to produce surpluses, an export market is
needed. The buyers in Europe are intelligent and are going to buy from
whom they can obtain regular quantity and quality the cheapest. This is
a partial answer to better farm incomes."[54]

Throughout the late 1960s, DeKalb County farmers were delighted to have markets for their surging output. It was one thing to benefit from international trade, however, and quite a different one to position DeKalb County farmers to participate directly in their own foreign marketing program. In the late 1960s, Mullins and others in the Bureau leadership proposed using their organization to create an international marketing business that would cut out the large agricultural shipping firms and thereby boost farm incomes in northern Illinois. Mullins turned to an experienced exporter, Sam Huey, for advice. In the spring of 1970, the pair discussed at length the potential for a Bureau-sponsored export company based in DeKalb. In June, Mullins met with the executive board of the Farm Bureau to discuss the plans for an exporting service. The key to profitability for such a small exporter was to charge a premium price for only the finest regional farm products and then use the higher prices to offset the transportation costs. During the summer of 1970, Mullins and Huey toured Europe and Japan to test foreign interest in DeKalb County produce and to drum up business. Their sales pitch was simple—the company would deliver higher quality grain directly to buyers who agreed to pay three to ten cents above market price per unit for that superior quality.

In September 1970, the pair made their first sale of DeKalb grain abroad, 2,000 tons of corn to Long John Distillers of Glasgow, Scotland. That autumn, they sold grain to a German company that was so impressed by the high quality of the 50,000 bushels they received from DeKalb that they paid a $3,000 bonus. Buyers in Japan were equally pleased by the quality of their trial shipments. The Farm Bureau board was excited by these successful trials and the potential of marketing DeKalb produce directly to discriminating overseas buyers. Events then moved quickly. In October 1970, the Bureau's executive board authorized two Bureau appropriations, totaling $185,000, for research and development of DeKalb County Exports, Inc. ("DCX"). The DCX board of directors was drawn straight from the leadership of the Bureau, the Bureau was the new company's only shareholder, and the original DCX headquarters was in the Farm Bureau building in DeKalb.[55] The Bureau hoped that the new firm, while focusing most of its attention on foreign sales, would also expand domestic markets, and in November Mullins and Bureau executive secretary Melvin "Mike" Hayenga traveled to Florida to look for new domestic customers. Bureau leaders were confident that their new company, backed by the finances of the DeKalb County Farm Bureau, could successfully compete with the major farm export companies in the world.

Buyers were clearly impressed by the premium quality and the specialty nature of DCX's offerings. Japanese brokers, for example, wanted select varieties of corn and food-grade soybeans, both commonly grown on DeKalb farms. By the spring of 1971, Huey had placed enough orders that DCX needed a large, reliable shipper to carry the DeKalb grain its customers. DCX began by shipping grain in sealed shipping containers, but that quickly proved unsatisfactory. Containers were expensive, required more handling than bulk grain-handling operations, occasionally leaked, and simply could not handle the amount of grain needed to fill DCX's orders. Huey turned to the Pillsbury Company, which already had an extensive shipping infrastructure in place, to subcontract the shipping. Huey also had previous experience working with Pillsbury. Pillsbury agreed to handle DeKalb grain for five cents per bushel. Success, however, brought problems. In addition to the high quality of its product, DCX also promised its customers prompt delivery, but Pillsbury declined to carry DCX grains after the first year of the agreement, probably because a crush of grain swamped elevator capacity nationally, especially in the Midwest. Pillsbury had enough grain to fill its shipping capacity without turning to the upstart DCX. A shortage of barges and loading facilities also threatened DCX's flood of contracts. Moreover, paying another company to handle DCX grain cut into farmers' profits and left the new company's success in the hands of others, a position that made historically independent Farm Bureau members and DCX executives uncomfortable. During 1972, the DCX board of directors outlined a bold new proposal—if DCX could not find suitable export partners, the new corporation would build its own shipping facility.

The decision to expand from a grain-marketing organization to actually owning a storage and loading facility was a major step that had significant long-term repercussions. The initial outline called for constructing a simple facility for handling corn and soybeans at a relatively modest cost of about $630,000. The recommendation got a lukewarm response from the Farm Bureau leadership. The DCX board reconsidered its plan, consulted with large grain exporters, including Cargill, and developed a more detailed proposal for a large, modern elevator and loading facility that would cost between $2 million and $3 million, to be financed by loans guaranteed by the deep pockets of the Farm Bureau.[56] The fact that the Bureau was the only stockholder of DCX and Bureau president Mullins was also the president of DCX made it an easy sell. The Bureau board of directors gave the project its blessing. Perhaps a more informative measure of the level of Bureau commitment to the undertaking was that it liquidated a portion of its cherished DeKalb County Agricultural Association stock and put the cash into short-term, high-yield securities to have funds readily available for DCX needs.

In the meantime, DCX representatives continued to tour the world, making sales contacts. During 1973, George Tindall, Ralph Montgomery, Don Huftalin, Bob Hutcheson, Don Peasley, and Mike Hayenga attended a U.S. Department of Agriculture food trade show in Tokyo, where they met with Japanese buyers. They went with high expectations. The post-war Japanese economy was healthy, and Japanese consumers were interested in more variety in their diets, including more red meats. The Japanese government was interested in increasing imports of American agricultural products to improve its balance-of-trade relationship with the United States. Moreover, the costs of domestically produced Japanese foods was extremely high, making American imports even more attractive. The market potential for high-quality grains and meat, both of which DeKalb farmers produced in abundance, seemed endless. Several men also visited a local hog farm to learn about Japanese techniques and better understand their feed-grain needs. By all estimates, the trip was a success. "Japan," wrote Don Peasley on his return, "can be one 'land of the future' for farmers." Asian inquires flooded into the DCX office—one asked about the possibility of shipping several million bushels of corn immediately, and Hong Kong brokers asked about the possibility of shipping several thousand dozens of eggs weekly.[57]

While Tindall and others were in Japan, Mullins toured Europe and the Soviet Union as part of a wider trade mission that included American economists, seed growers, and agricultural processors. In contrast to the Asians, Europeans were cooler toward DeKalb products. The Soviet Union chronically underproduced, and prospects for internal improvements in farm efficiency and better yields made it an unlikely market. At the other extreme, Western European nations were modernizing their own agricultural systems, practicing an intensive agriculture that improved yields for domestic consumption. European governments heavily subsidized their farmers and used tariffs to protect domestic producers. Perhaps the biggest obstacle for DCX sales representatives on both continents, however, was the U.S. reputation for inferior exports. "[European] buyers," wrote Mullins, "told the visitors from the U.S. they'd buy corn from six other countries before turning to the U.S. 'because your country has a reputation for poor quality.'" It was sometimes an uphill battle convincing European buyers that DCX's business philosophy was different and that its grains commanded premium prices because they were superior quality. The greatest interest in DeKalb products, therefore, came from customers, like food processors and distillers, who required better grades of grain for human consumption. During 1973, DCX exported about 1.5 million bushels, and orders were arriving faster than it could fill them. The company was already valued at $1.32 million and had sufficient assets to cover all of its liabilities. By the next summer, the DCX board of directors' goal was to ship 10 million bushels of specialty corn and 2 million bushels of premium soybeans annually. The future looked promising.[58]

The flood of orders and a regional shortage of storage and transportation capacity underscored the need to complete the DCX high-speed loading terminal as quickly as possible. After an exhaustive search, the DCX board of directors settled on a forty-acre former coal strip mine along the Illinois River in Ottawa, Illinois, as the location for its new facility. The site was hardly ideal—contractors needed to complete extensive stabilization work on the wasteland before construction could begin—but it had two overriding advantages: it had 1,400 feet of river frontage, and it was close enough to DeKalb farms to keep transportation costs to a minimum. Groundbreaking for the first phase took place in October 1973, just days after final approval for the project from the U.S. Department of Interior. By the following June, 1974, four dolphins for anchoring barges had been installed, the barge-loading tower was nearly complete, the skeleton of the elevator itself was in place, and the silos and truck dump pits were roughed-in. By late summer, the terminal was ready to load barges and the entire facility was on target for its mid-1975 completion.[59]

While construction continued, DCX was perfecting its marketing plan. DCX hired Jim Duvall to replace Huey as manager and retained Huey as its sales representative in Europe and Asia. It also hired Dale Hamilton, a DeKalb County native, as its domestic purchasing representative. Hamilton's job was particularly important. The core of the DCX business philosophy was to offer customized merchandizing, based on guaranteed grade, protein content, and oil content, to overseas buyers. In turn, DCX offered area farmers six to ten cents a bushel premium on corn, based on DCX's own rigid standards. DCX was also looking to capitalize on the promising world market for soybeans and encouraged farmers to raise high-quality soybeans for human and animal consumption that would attract international buyers. "These are standards almost every farmer can meet with careful handling, thoughtful storage and control over his crop," wrote Duvall. "What we are doing is making him aware of this profit and providing him with the means to receive a price nearer what his corn and soybeans are worth." With the facilities and the management team in place, DCX looked forward to becoming an important worldwide commodities broker.[60]

In the autumn of 1974, the first phase of construction was complete with the installation of the pier, conveyer belt, grain cleaning and testing equipment, truck scales and dumping equipment, and dust-control devices. The first load of corn went from the truck dumps, up the conveyer belt, and onto a barge for overseas shipment on December 2, 1974. The total cost projections, however, had been revised upward to over $3.6 million, somewhat more than the original estimates of between $2 million and $3 million, and the terminal was still not completely finished. Cost overruns, however, attracted little attention. Export estimates and orders suggested DCX might ship as many as 2.5 million bushels during 1974—in fact, DCX handled 3.8 million bushels during the first six months of the 1974–1975 fiscal year, a substantial increase over the volume of the previous year, when DCX had no loading facility of its own. When the terminal was fully operational, DCX anticipated shipping maybe ten times as much. The future looked bright, and, in the wake of such favorable business predictions, the company pressed ahead with its new building despite the skyrocketing construction costs.[61]

In addition to working familiar territory in Asia and Europe, the Bureau sent representatives on trade missions to the Middle East, including Lebanon and Iran, where high oil profits, increasing populations, and lack of water and arable land made them seem to be perfect candidates for DCX's direct-marketing approach. Foreign interest in American processed meat, especially in Japan, also caught the attention of the Farm Bureau's marketing committee, which began to explore the possibility of exporting beef and pork to Asia. Such trade missions proved successful, and in 1974 DCX introduced a marketing program for ordinary grain in addition to its trademark premium-quality and select variety offerings.

All phases of the DCX facility were finally completed in the summer of 1975. Cost overruns pushed the final capital investment in DCX to about $5.2 million, but the DCX elevator was the first new terminal built on the Illinois River in decades and was a showplace for modern agribusiness. Extensive grain-cleaning equipment helped ensure the highest quality. Fast conveyor belts loaded 25,000 bushels per hour into barges. Huge dumping lifts tilted semi-tractor rigs at a steep angle to pour out grain at a rate of one trailer every two and a half minutes. Dust control ventilation kept the terminal safe, and pollution abatement measures ensured the facility was environmentally sound. By the fall of 1976, the elevator had a modern Berico grain-drying unit, the largest in the area, capable of extracting moisture from 7,200 bushels per hour, and a new inspection house where loads were precisely sampled and graded before reaching the dump pits. Sensitive temperature-monitoring systems reduced the possibility of spoilage within the elevators. During 1975, the terminal handled about 14.5 million bushels of corn and soybeans. During 1976, DCX decided to handle "ordinary grain" business in addition to its specialty grain exports. That year, it handled more than 23 million bushels of ordinary and specialty grain. Business volume seemed to fulfill all the optimistic projections.[62]

Yet, despite the seemingly favorable statistics, DCX never lived up to expectations. Although it shipped large amounts of grain, it was never enough to realize a profit. It lost money steadily, but the Farm Bureau continued to back it, helping the enterprise maintain cash flows and prevent bankruptcy. One symptom of the problems was fast turnover in leadership during the company's formative years. Although the core of officers in DCX remained fairly constant, with George Tindall taking over from Mullins as president, Ray Willrett as vice president (later president), and Paul Montavon as secretary, the critical position of manager was in flux. Duvall had already replaced Huey as manager of DCX in 1974, but he left in October 1975. The DCX board hired Leo Smith, a grain merchandiser, as interim manager and then hired Eugene Bissell as manager in February 1976. Huey also quit as DCX salesman and distanced himself from the enterprise in March 1975, taking with him a wealth of experience and contacts in international agricultural marketing, shipping, and finance. In March 1979, Bissell resigned, and DCX hired Kevin Heupel as general manager.

To make matters even more uncertain, Mullins himself retired as Farm Bureau president in February 1974, after eighteen years of service. Mullins, who had been instrumental in the creation of DCX and one of its most persuasive supporters, as well as a leading voice in the DeKalb County Farm Bureau, also resigned from the DCX board of directors two years later along with other key DCX supporters, Richard Walter and Ralph Montgomery. The new Farm Bureau president, Donald Chaplin, took over during troubled times for American agriculture. Farm production costs were escalating, and farm incomes remained stubbornly low despite the rosy economic projections offered just a few years earlier. Average farm incomes, based on a 1910–1914 index basis, were the lowest they had been since the Great Depression. Chaplin supported DCX and saw it as the best means to make DeKalb farmers competitive in the world market. But, although the new DCX facility was shipping considerable amounts of grain and paying premium prices for select varieties and quality, local farmers were still finding it difficult to make a profit and the company was losing money. Under the circumstances, the costs associated with operating DCX increasingly seemed like an indulgence. Bureau members began to question the need for their own elevator, especially one that, at more than $5 million in final cost, was nearly double original estimates.[63]

Despite the internal problems at DCX, throughout the late 1970s it took contracts from foreign companies that appreciated the high-quality grain that was the DCX trademark. Not only did DCX guarantee the grade, but it also watched carefully as it was loaded into shipping containers or barges to ensure that the grain kernels did not shatter during the process. Such damage usually rendered the corn inferior or unusable for most high-grade applications. The new Ottawa loading facility was designed specifically with that careful handling in mind, and DCX representatives often oversaw the transfer of grain cargos from one ship to another or onto docks, railroad cars, or trucks. DCX relations with its customers were outstanding, and contracts flowed into DeKalb. One Scotch whisky distiller, for example, was delighted with the virtual door-to-door service and, more important, found that they could distil 15 percent more whisky from the carefully graded and handled DeKalb corn than they could from corn shipped by other companies.

At home DCX relations with area farmers were not as smooth. DCX offered a premium for acceptable grain, and most farmers were happy to deliver the grades and varieties they promised. A few, however, mixed inferior grades into their DCX deliveries. Others handled their grain so poorly at harvest that it was unfit by the time it was delivered abroad. Buyers paid less for such adulterated and damaged grain or else refused it outright, cancelled their DCX contracts, and turned to other grain suppliers. DCX was sometimes scrambling to purchase expensive grain on short notice, at inflated prices, to fill its customers' orders. Although DCX probably had sufficient contracts to make a profit, DCX found it difficult to fill them profitably.

Soaring energy costs, double-digit inflation, soft stock market returns on the Bureau's portfolio, and high interest rates also undercut DCX's ability to remain competitive. The Farm Bureau, which relied heavily on its investments, including its DeKalb County Agriculture stock, to fund its programs, was hit hard by the troubled economics of the 1970s. Some Bureau leaders, and probably a majority of farmer members, were having serious second thoughts about the scope of the DCX undertaking and did not hesitate to voice their concerns at Farm Bureau meetings. Livestock farmers were especially upset that the Bureau was focusing so much attention on grain marketing and wanted it to make a similar commitment to marketing beef and pork. Many members feared that the Bureau was incurring too much debt, paying out too much cash, and, in general, straying too deeply into the grain business at the expense of its core functions. Support for the whole international marketing venture began to evaporate as the risks mounted and the costs increased.

Finally, events seemed to conspire against DCX success. In 1977, the Illinois River froze for only the second time in recorded history. Barge traffic ground to halt, stopping DCX shipments for two months. Nearly 34,000 bushels of corn sat spoiling in uncovered piles on the ground because the river was closed and the elevator was full. Later that year, the Army Corps of Engineers closed the river to barge traffic for another two months to repair damaged locks. The costs associated with the delays and the damage to the grain stored outside cut deeply into DCX's operating funds. Another problem was land access to the facility. From the beginning of the project, Canal Road, the access road to the terminal, was incapable of handling the traffic of heavy grain trucks. In 1978 a series of problems with the road prompted the DCX board to finally negotiate with state and local authorities to develop a scheme to share responsibility, and costs, to rebuild the road. A malfunction in one of the grain dryers in 1979 triggered a serious fire that resulted in the complete destruction of one dryer and damage to the second unit, causing loss of drying reserve and more delays. Larger multinational grain companies, which had a broader customer base and product selection, were in a better financial position to weather economic losses and reversals. DCX was too small to compete with the big companies, and there was a limit to what its discriminating clientele would pay for DeKalb grain. DCX simply could not benefit from the economies of scale that kept the big companies profitable during volatile times for grain exporters.[64]

Despite growing sales, by the early 1980s, it was clear to many in the Farm Bureau that DCX specialty grain business was in serious financial trouble and the Farm Bureau was finding it increasingly difficult to support the project.[65] The final blow came in 1982. The grain business had always been a highly speculative one, but the DCX board, as well as Farm Bureau leaders, always cautioned DCX managers against speculating too heavily in grain futures. Specific requirements limited managers to hedging only small lots of grain for short times. No one at DCX or the Bureau, however, thought about limiting managers' abilities to speculate in barge transportation capacity futures. Heupel speculated heavily on barges on the Illinois River, hoping to cut DCX costs by securing carrying capacity when prices were low and boost profits by selling any excess capacity on the open market. Heupel, however, guessed wrong, and the price of shipping declined steadily. To make matters worse, he continued to believe the prices would recover and ignored advice from brokers to cut DCX's losses and get out of the downward cycle. It proved to be a huge financial miscalculation that cost DCX and the Bureau millions of dollars. The Bureau was on the verge of not having enough cash to cover all its debts. In the autumn of 1982, the Bureau board considered its options during a tense meeting. One message was clear: the losses at DCX were so great, and future prospects for profits appeared so dismal, that DCX needed to be sold quickly. Allan Aves, who took over from Chaplin as Bureau president in 1977, and the Bureau's new business manager, Doug Dashner, who began at the Bureau in early 1979, took it upon themselves to locate a buyer for DCX's assets and prevent a financial collapse that threatened to permanently impair, or even destroy, the Farm Bureau itself. It was difficult to find a buyer, but the Bureau eventually sold DCX and its new terminal facility to Archer Daniels Midland's Tabor Grain division for $450,000, about a 90 percent discount over the terminal's final construction cost and a nearly $4.5 million loss for the Farm Bureau. ADM/Tabor agreed to continue to accept regular grains, contract for the specialty grains many DeKalb farmers already owned or had planned on growing, honor DCX's existing contracts, and, most important, assume all DCX debts. After eight years, the Bureau was officially out of the grain-marketing business, much to the relief of most Bureau members. In a most ironic postscript, shortly after the Bureau sold DCX to ADM, the price for carrying capacity on the Illinois River rose three-quarters of a cent per bushel, an amount that would likely have resolved the short-term barge speculation problem.[66]

After the sale, it was popular to blame the large grain-marketing companies for DCX's troubles. "They just squashed us, you know," Aves said years later, "and there was nothing we could do about it." There is no doubt that the grain business during the late 1970s was highly competitive and that large companies were in a better position than DCX to take advantage of the volatile market. In retrospect, however, it is easy to see that a variety of events led to DCX's downfall. More far-sighted visionaries might have developed a different market strategy. A board of directors composed of agricultural business experts, instead of farmers who were talented business amateurs, might have anticipated and solved problems earlier. A more financially conservative DCX and Bureau leadership might have raised tougher questions about cost overruns and regular DCX losses. Area farmers who blended grain and lowered its quality might have profited more by honestly delivering grain and collecting payment premiums. More astute terminal managers might have better handled market downturns and shipping bottlenecks. A frugal financial manager, as Dashner proved to be, perhaps would have been more skeptical about optimistic profit projections.[67]

Perhaps the greatest shortcoming, however, was so deeply imbedded in the mind-set of Bureau leaders that it would have been nearly impossible for them to recognize it. DCX was, after all, another in a series of imaginative DeKalb County Farm Bureau experiments intended to boost the financial fortunes of area farmers. And, just as the Bureau itself was founded on ideals that were at the cutting-edge of agricultural innovation in 1912, the DCX marketing philosophy was ahead of its time. A decade or two later, when American and international buyers became more keenly aware of the quality of their foods, a small, specialty grain–marketing venture like DCX may have been more successful. Farmers during the succeeding decades came to better understand and appreciate the long-term value of raising grain for niche markets. In the end, however, the Bureau invested millions in the project, and, while for eight years it offered area farmers premium prices for their crops, the financial benefits reaped by individual farmers came at a huge collective cost to the Bureau and its membership. DCX, which began as a plan to help make DeKalb grain producers competitive in the global food market, proved to be the most serious financial crisis the Bureau ever faced.[68]

The Bureau's financial bind during the early 1980s was exacerbated by its other experiment in collective marketing. For decades, since the heyday of the Bureau's locker service in the 1930s, the Bureau used its purchasing power to offer members food products at reduced cost. Originally it sold mostly frozen fruit, but the service proved so popular that eventually the Bureau offered a whole variety of items and formed a separate marketing committee to coordinate the sale of citrus, nuts, special varieties of dairy products, orange juice and lemonade concentrate, meat, popcorn, fresh vegetables and fruit, and miscellaneous seasonal treats to Bureau members. The marketing committee also took advantage of the Illinois Agricultural Association direct-marketing program to bring lower-cost fresh and processed food to Bureau members. During the 1970s, while the Bureau was focused on DCX, area stockmen asked the Bureau for a similar marketing service for their meats. The trade missions and sales trips to Europe and Asia suggested that there were foreign markets for DeKalb beef and pork products. Domestic buyers might be attracted to the high quality of meat produced in the county. The future for a direct meat sales business looked bright, and, in 1976, the Bureau created a new, wholly owned subsidiary company, DeKalb County Marketing Services, Inc. (DCMS).[69] Although the main emphasis of the new company was marketing meat, DCMS also sold fruit, vegetables, cheese, and other products to DeKalb County residents. DCMS acquired many of the items it sold from farm bureau cooperatives in other states, and, likewise, DCMS sought to market its "Country Pride" meat to farm bureaus across the nation. The idea was simple—provide a system for direct farmer-to-consumer sales for meats. By keeping processing costs to a minimum, using its purchasing power to negotiate lower prices from suppliers, and cutting out the costs associated with shipping, handling, processing, and the like, DCMS's direct-marketing plan promised lower prices for consumers and higher returns to area farmers.

DCMS expanded quickly. During its first year, the company opened the doors of its retail operation at the Farm Bureau building and installed freezers and display spaces. It aggressively pursued marketing opportunities, especially with the Bureau's petroleum affiliate, Northern FS, Inc., which it hoped would carry Country Pride products in its stores. The marketing committee members hit the road looking for deals to bring back to DeKalb. DCMS officials consulted with Swift and Co. and DCX about the potential for shipping meat overseas and was especially interested in the Japanese market. In 1978, the Bureau bought a building on North Sixth Street, in DeKalb, for the DCMS headquarters, store, and warehouse, and hired Mark Girardin as business manager.

A year later, DCMS was showing a steadily increasing volume of sales for all its offerings and had even greater ambitions to expand the meat-marketing program to compete with regional, and even national, meat suppliers. Between 1976 and early 1979, DCMS obtained meat from a handful of small regional slaughterhouses that met the high DCMS standards. In February, 1979, with the DCMS board aggressively shifting the company's emphasis toward its meat-marketing service, DCMS's primary meat supplier, the Earlville meat locker owned by Bert Halsey, was destroyed by fire.[70] Other area meat packers could not produce the quantity or quality of meat that DCMS demanded for its expanded marketing goals, and the large meatpacking firms, like Swift, were not interested in filling the comparatively small DCMS orders. DCMS and Farm Bureau leaders, however, were undeterred. A series of feasibility studies commissioned by the Farm Bureau and DCMS boards showed that a new meat-processing plant could successfully provide four basic services: custom slaughtering for local customers, a retail shop for fresh and frozen meat sales to consumers, a direct-marketing program to other farm bureaus and national and international customers, and a wholesale business to sell meat to restaurants and other food service providers. A new slaughter plant seemed to be the best means for providing quality meat for this enlarged meat-marketing program while assuring stockmen fair prices and consumers top-quality meat. "We wanted to say, 'we believe in the livestock industry,'" reflected Farm Bureau business manager Doug Dashner.[71]

In September 1979, the Farm Bureau and DCMS made good on that belief and announced that DCMS would build its own livestock holding, slaughtering, and packing facility capable of processing at least 600 head each month.[72] "We realized, finally," said Dashner, "we would have to either build our own place or get out of the [meat] business altogether." "[T]he ability to have quality meat available and to increase meat consumption are the reasons we're behind it," said Bureau president Aves. "I don't really have any great concerns. I'm certain it will work." DCMS board member John Huftalin said simply, "Need prompted us to make this decision. It's an ambitious decision. Nobody had done anything quite like this before." Construction began the next month on a site about one and a half miles south of DeKalb, and the new building was completed a year later at a total cost, including overruns, of $1.5 million.[73]

Having the Farm Bureau involved in a slaughtering business was not new—the old Locker Service, after all, offered custom packing. Indeed, the Bureau leadership activity sought to characterize the new enterprise as a continuation of the Bureau's traditional service ideals. What was different from earlier efforts, however, was the scope of the enterprise. "This facility will revive that old-fashioned concept of providing service that the customers ask for and deserve," said Aves at the Country Pride Meats grand public opening. "The facility," he continued, "is a return to the old-fashioned concepts that made agriculture the strong backbone of our country, our state, and our nation. The notable difference is that these old-fashioned concepts are being delivered by the most

modern technology available today." Even the name of the new enterprise, "Country Pride" evoked images of rural strength and values.
"The Farm Bureau," noted Aves, "takes 'pride' in Country Pride. We
hope the community will also take 'pride' in this facility and what's
even more important, 'pride' in the quality meat that's produced right
here in DeKalb County."[74]

DCMS in many ways embraced similar ideals and business plans of
DCX. The DeKalb County Farm Bureau was the first farm bureau in the
nation with a complete producer-to-consumer meat-marketing program.
Like the DCX elevator, the new DCMS facility was state of the art, was
energy efficient, and matched or exceeded the highest standards of the
industry for safety, sanitation, and health. It had eighteen full-time employees and was, said manager Girardin, "the Cadillac of the meat industry."[75] The DCMS sales plan reflected the DCX ideal and rested on a
philosophy of providing superior products, including fresh and frozen
cuts of red meat and poultry, cured and smoked meats, and a variety of
prepared delicatessen products. The Bureau also expected to reduce middlemen and operating costs, eschew profits, and return more money to
area stockmen. The majority of the livestock would be purchased from
Bureau members, who would receive premium prices for superior-quality
animals. DCMS sales representatives explored overseas opportunities
and stepped-up efforts to expand domestic sales beyond northern Illinois. The company, for example, opened direct-marketing pilot programs for DeKalb meats in Chicago, neighboring Illinois farm bureaus,
out-of-state bureaus in the Midwest and the South, and independent
meat suppliers in Illinois and as far away as Florida. By the end of 1980,
DCMS, led by its flagship Country Style Meats division, reported a 40
percent increase in sales.[76] It was the fourth-largest direct-marketing operation in the nation and anticipated becoming the second largest by
the end of 1981. Supported by the deep pockets of the Farm Bureau and
the Bureau's leaders and a majority of members, the possibilities for
DCMS seemed limitless.[77]

Similarities with DCX extended beyond the realm of business philosophy and goals, however. As was the case with DCX, the DCMS vision
failed to materialize. The market studies completed in the late 1970s
proved to be overly optimistic. There were never as many direct sales as
predicted. Retail competition from chain grocery stores undercut consumer sales. Although DCMS tried to convince consumers about the di-
etary benefits of meat, Americans in the 1980s were changing their buying habits and eating less meat. Consolidation of the livestock industry

meant that fewer farms raised livestock and resulted in less demand for custom slaughtering services for home consumption. The 12,500-square-foot processing plant handled a considerable number of orders, but it seldom operated at full capacity. It was a losing venture from the beginning, and the profit horizon was too long for it to survive. Large national and regional meatpacking companies operated with economies of scale that DCMS found impossible to match. Consequently, Country Style products, while often of superior quality, were not competitive with national brands.

At the same time, the DCX failure cut deeply into the Bureau's pocketbook, and many Bureau leaders feared that another significant financial reversal would threaten the very existence of the Bureau itself. Moreover, the DCX experience lead many in the Bureau to reconsider the Bureau's core mission. It was acceptable to have the Bureau affiliated with traditional service enterprises, like Country Companies insurance and the Northern FS petroleum and farm supply business. These successful ventures, after all, began as modest operations that grew steadily over the decades. While the Farm Bureau had an interest in their success, it was not the sole stockholder of either one, as it was for DCX and DCMS. Even the slow decline of earlier Farm Bureau–affiliated operations, like the old Locker Service, were not much of a setback for the Bureau compared with the demise of DCX and DCMS and the loss of millions of dollars of Bureau assets. Bureau members increasingly came to believe that it had been a mistake to venture into wholly owned businesses on such a scale, especially when Bureau leaders had little agribusiness background. In both cases, after all, the boards of directors of each company consisted mostly of Bureau leaders who were certainly successful farmers and stockmen, but hardly had significant experience in the national and international business worlds they sought to dominate. Talented amateurs were capable of founding and operating one of the most successful farm bureaus in the nation, but marketing produce to the world was a job for professionals. In 1987, DCMS closed its operations and sold its new processing facility, and the Bureau abandoned the meatpacking business.

The DCX and DCMS ventures represented the high tide of DeKalb County Farm Bureau commercial activities and marked the limits of what one of the most powerful and influential farm bureaus in the nation could accomplish. In the rough-and-tumble world of agribusiness, even the best-intentioned ventures fall victim to harsh market realities. Moreover, management of the companies was largely in the hands of

Farm Bureau leaders who were solid farmers but lacked professional experience in agricultural marketing on the scale required to make the Bureau's companies successful. The failures of the businesses also demonstrated that, despite the Bureau's deep pockets, the organization was essentially a conservative one that was too risk-averse to stake everything, including its very existence, on its commercial ventures. The Bureau, after all, was not a corporation, and even at the height of its marketing ventures it never departed from its original mission of providing practical assistance to its members. High-stakes business goals and Farm Bureau member services were never as compatible and mutually complementary as Bureau leaders believed, and in the end the Bureau remained true to its core values.

Historically, the DeKalb County Farm Bureau had been a local, or at most a regional, organization dedicated to assisting its members cope with the new and rapidly changing farming conditions in the twentieth century. It provided solutions to problems that were largely beyond the control of ordinary farmers. It regularly supported, for example, policy measures at the local and state levels that were of particular concern to farmers. Likewise, the Bureau from time to time arranged marketing pools that helped producers compete more effectively in regional markets. In this light, the Bureau's sponsorship of the Farm Policy Council and its attempts to enter the national and international agricultural markets was a change in magnitude, but not in kind, of the sort of assistance it had offered its members for decades. When farmers in DeKalb County were confronted by political and economic conditions that impaired their livelihoods and threatened their financial futures, it seemed natural for the Bureau to attempt to remedy those conditions by whatever means seemed practical, including venturing beyond the boundaries of its traditional ideals. One way of attempting to influence the conditions under which farmers labored was by reshaping the national political landscape; another was by setting new standards for agricultural marketing. If the Farm Policy Council sought to change the policy landscape, the Bureau's businesses sought to attack the farm problems from a marketing standpoint. In the end, the policy initiative proved successful; the marketing companies did not. Perhaps the most remarkable feature of these programs was that the Bureau leadership saw them not only as possible, but absolutely essential for the very survival of agriculture in the county. Later in life, Howard Mullins reflected simply, "I had to do it [the Farm Policy Council] so that I would have something to leave to my children."[78]

The demise of DCX and DCMS signaled the end of an era for the DeKalb County Farm Bureau. The turbulent post–World War II decades witnessed a fundamental realignment in American agriculture that had enormous influence on DeKalb County farmers. New technologies and market conditions drove farmers to produce at levels that eclipsed even the most productive years in the first half of the century. Farmers in the period faced challenges, both unique and familiar, that tested their resolve. The Farm Bureau helped them cope by offering an array of important traditional and new services. Yet the same energy that motivated the DeKalb County Farm Bureau to so diligently support its members also drove it into unfamiliar territory in the name of providing that support. In the closing decades of the twentieth century, the Farm Bureau returned to its roots to help area farmers confront a new range of problems associated with modern agriculture. The financial reversals of the 1970s and 1980s, a crisis of confidence resulting from the DCX and DCMS failures, the declining farmer membership in the Bureau, and the demographic profile of the remaining Bureau members at the turn of the century, however, would mean that the Bureau's programs would be very different as the organization also confronted its own limitations.

6 THE FUTURE

• On September 12, 1996, the DeKalb County Farm Bureau dedicated its new Center for Agriculture, its fourth home since 1912, at the corner of Bethany and Peace roads in Sycamore.[1] Local dignitaries, Bureau officers and members, and members of the public attended the event—more than a thousand people toured the $6.2 million facility in the first two days after it opened. Congratulatory notes arrived from national leaders. "We dedicate this facility," said president Ken Barshinger, "in the memory of those who have gone before us, to be used productively by those who are among us, and with a vision to the future of those who are yet to come." Despite the crowds of well-wishers, the gala lacked the innocent exuberance that marked the dedication of the Bureau's offices nearly fifty years earlier. Then, Bureau leaders were confident that their organization would guide DeKalb farmers toward an inevitably bright tomorrow. "Today, our voice is heard in the state and national councils," wrote William Eckhardt in 1951. "Our strength will become the greatest of any organized group, and it is my hope that this power will be used with moderation and wisdom." Henry Parke charged his beloved Bureau to "lead ever to higher standards and continue to strive for peace, preserve our freedoms, and strengthen our democracy." In late summer of 1996, such faith seemed naive. The farm bureau movement at the close of the twentieth century was no longer the "great voice for American agriculture" it had been at midcentury. Assumptions about the fundamental importance of local farm bureaus as the embodiment the American agricultural legacy and cold war bulwarks against communism were long forgotten. Nobody called the new Center for Agriculture, as they had the earlier Bureau office, a "temple of agriculture," and few still seriously believed that Americans viewed farming as the "highest calling of our nation." The DeKalb County Farm Bureau remained independent and active, but across Illinois and the nation, bureaus were consolidating with neighboring counties or simply closing their doors for lack of interest and money. The subdued tone is partly explained by the internal controversy among members about whether to build the new office when the Bureau had so recently suffered serious financial reversals with the DCX and DCMS projects and seemed generally less financially secure than it had been during the three decades following World War II. More significantly, however, the celebration reflected a much more mature, realistic, and somber attitude about the future of agriculture in DeKalb County—a future that was far more uncertain at the end of the twentieth century than it appeared to be at midcentury.[2]

The star of the event was the 40,000-square-foot building itself, designed by the architectural firm of Larson & Darby. Everything about the design of the red brick building evokes agricultural themes. High and steeply pitched rooflines rising from the surrounding fields resemble barns. The broad footprint of the structure keeps it firmly rooted to the ground. The Bureau's round amphitheater reminds visitors of the hundreds of corncribs, grain elevators, and silos that are ubiquitous throughout the county. Natural fieldstone accents make it appear as though the building is emerging from the ivory colored limestone that is the foundation of the county. When it was completed, area residents looked out over ripening fields on fair days and saw the building framed between the green crops and the pale blue sky. Since that time, however, the open space that accented its most impressive design features is gradually being engulfed by development—the seemingly inevitable symptoms of urban sprawl. The visual bond the structure has with the land that surrounds it diminishes more with each passing year. Soon, the original design will be forced to compete with a new set of aesthetic values, or perhaps the absence of aesthetic values altogether, as the agricultural landscape is relentlessly transformed into a suburban one.[3] And yet what is happening to the Bureau's new home reflects even more difficult changes occurring within the organization itself. The DeKalb County Farm Bureau, like its new building, is not likely to disappear anytime in the near future. Changing social, economic, and political values, however, will require the Bureau to operate differently in the future, just as certainly as new ideas, values, and expectations reshape the agricultural landscape of the county.[4]

Inside the foyer of the Center for Agriculture, a massive carved wooden mural greets visitors. Local carver Joe Dillett's seven-by-thirty-two-foot artwork depicts changing seasonal and historical themes in

DeKalb agriculture. That story is so closely intertwined with the Farm Bureau during the twentieth century that the mural is equally a graphic rendition of the Bureau's history. During the more than ninety years since the Soil Improvement Association began, DeKalb has experienced three eras of significant social change that shaped America. The first era, described by rural social scientists as the "community control" period, lasted from the opening of the century until the outbreak of the Second World War. The second period, the era of "mass society," began in earnest at midcentury and lasted for about forty years. The mass society model, while dominant for decades, is fast declining in influence. In its place is a third, emerging era, the "information age," that is likely to dominate the organization of society well into the twenty-first century. The challenges that confronted rural America, and that it will face in the future, are not all unique, but they do reflect major forces that shape social organization throughout the world. The DeKalb County Farm Bureau in the coming decades must respond effectively to those issues.

During the community control era, communities were geographical locations where residents lived out their lives. Place influenced society. In DeKalb County, communities were most commonly the scattered small towns and villages. By the time of the Soil Improvement Association, the concept of community in the predominately agricultural DeKalb County had expanded somewhat to include the county itself, especially with the advent of improved roads, automobiles, telephones, radio, electricity, and all the wonders of the early twentieth century that made rural America more integrated and connected. Most folks remained closely tied to their community, both their local town and the county, which exerted powerful controlling forces on their daily affairs. The strength of community was especially evident in rural areas like DeKalb County, which had a small population with low density, great social homogeneity, and low population turnover.

Within rural communities there was also overlap of social institutions that helped provide structure to rural life. Community elites played important leadership roles in a variety of institutions and thereby exerted considerable influence over them and local society at large. The original Soil Improvement Association, for example, was a new and powerful community institution that was founded and led by men who exercised power, often simultaneously and in several towns within the county, as farmers, bankers, journalists, businessmen, politicians, educators, and churchmen. Likewise, there was an inherent assumption, especially among Progressive-era agricultural reformers in the pre–World War I period, that strong community institutions improved lives for citizens. President Theodore Roosevelt's Country Life Commission, for example, sought to secure a place for farmers in a rapidly changing American society of the time by relying on improved technology and institutional reforms to create modern, forward-looking rural communities. In the eyes of the commissioners, rural social welfare was inextricably linked with strong communities and economies. Likewise, the Soil Improvers in DeKalb created an organization dedicated to enriching local agricultural, commercial, and educational institutions, which they assumed would ultimately improve the lives of all DeKalb residents.

About a third of the American population, and two-thirds of rural Americans, still worked the land during this community control era. These people tended to view social issues in the context of farming not as a career, but rather as a way of life. The farmer, after all, owned the enterprise, provided the labor, managed the risks, made most operational decisions, and lived on the farm. Indeed, one of the great challenges the Soil Improvement Association and the Bureau faced during the community control era was to graft a modern education-based, technology-driven, business-oriented farming model onto the traditional agrarian lifestyle of DeKalb farmers to make them more economically competitive and socially successful. The issue for Eckhardt and organization leaders was never the availability of new ideas and technology—the twentieth century was perhaps the most agriculturally innovative period in the history of humanity—but how best to diffuse those new techniques throughout the countryside. In that regard, the consistency and cohesion of the DeKalb County community, the respect that people had for the leaders and advisors of Soil Improvement Association and the Farm Bureau, and all the interlaced social and institutional relationships within the community, made it easier for innovators such as Eckhardt, Roberts, and Bureau leaders to guide DeKalb farmers toward the future patiently.

At midcentury, tightly knit rural communities confronted a series of new forces. The same things that expanded the sense of community to include all of DeKalb County—especially improved communications, better roads, faster cars and trucks, and expanding population—now threatened to unravel the traditional stability of the community and its institutions. Military service and higher education brought young rural folk into contact with new places, people, and ideas. Greater disposable wealth gave their parents more opportunities to travel outside their customary spheres and experience similar things. New engines of economic development moved into historically rural areas and brought with them nonfarm jobs and outsiders. Metropolitan areas and suburbs grew and expanded their influence into rural areas. National, and even international, corporations began to compete with local and regional providers of products and services. The complexity of new agricultural technologies meant that farmers relied upon outside experts, such as scientists and technicians who had little personal investment in individual communities, to provide expertise in farming matters. New agricultural methods also forced farmers to look to outside suppliers for the necessities of agriculture, such as hybrid seed, chemicals, and specialized machines. The entry of women into the labor force supplanted old social networks with new ones that were not necessarily centered in traditional community institutions. Industrial and professional careers, coupled with dwindling farming opportunities, attracted rural youth away from communities, further eroding historical community ties. Regional and national chain retailers competed with local merchants. Even more than radio had a generation earlier, television homogenized the national population, beaming the same programming and advertising to urban dandies and rural hayseeds. In time, more American households had a television than had indoor plumbing, and the networks and advertisers standardized American tastes and expectations. Some small towns evolved into urban commercial centers, while others withered away until only a few homes and weathered storefronts marked a local community upon which its residents had depended for decades.

The influences that marked the community control era gave way under such pressures to the era of mass society, which first appeared after World War I but came to dominate American society at the end of the Second World War. Communities as the primary spheres of social interaction opened more and more to outside influences. Important matters that had profound local consequences, such as civil rights and environmentalism, were increasingly likely to be decided beyond community boundaries, and often with little regard to how those decisions might ultimately affect local social relationships and institutions. There was also an emphasis in the mass society period on developing new and increasingly large hierarchies of power. Local representatives of product and service providers, for example, answered to regional and national corporate headquarters. The federal government increasingly exercised power over daily life of farmers at the expense of state and local governments. Agricultural or foreign policy decisions made in Washington could have profound impact on DeKalb farmers, while the power of farmers to influence that policy through the traditional operation of representative government diminished as the percentage of farmers in the national population declined. In response, farmers turned to their farm organizations in an effort to amplify their voices in government circles. Local farm organizations became bound even more tightly to their national counterparts. State agricultural associations and the American Farm Bureau Federation, for example, sought to become the collective voices of American farmers in the halls of political power. The DeKalb County Farm Bureau was an integral, though sometimes reluctant, partner in the state and federal organizations but was also powerful enough at one time to wield power at the national level independently, as in the case of the Farm Policy Council.

For much of rural America, the mass society era provided mixed blessings. Decades of agricultural changes—such as reform efforts, government support, new technologies, increases in societal wealth, and marketing success—brought many farmers into mainstream society. Rural folk came to believe that improved agricultural technology and markets would enable them to operate larger farms, be more productive, and earn more disposable income. For the first time, American farmers came to expect that their lives should be the equal of successful urban and suburban ones—modern homes, modest luxuries, vacations, new cars, economic opportunity, advanced education, health care—that were illustrated so pervasively by the new mass communication, especially television. Indeed, in many places, like fairly prosperous DeKalb County, the quality of farm life was on par with urban life in most respects.

On the other hand, the benefits brought about by modern agriculture went to fewer and fewer farmers. Each year, more and more farmers gave up farming and opportunities to enter farming as a career diminished. Although this process reduced political power and social influence of farmers in the long run, rising prosperity made daily lives of farmers who remained much better. Rural children might have been deprived the opportunity to follow their parents into agriculture, but they found new opportunities and material success in other careers. The mass society period also witnessed increasing specialization of agriculture. DeKalb County, for example, was typical of most corn belt counties. At midcentury, area farms raised a variety of products. While most of the county's land was devoted to corn, many farmers also raised some slaughter livestock, poultry, and dairy cattle—indeed, for much of the early decades of the twentieth century, DeKalb County agricultural income was derived as much from livestock as from grain. Until well into the 1970s, it was not uncommon to see large enclosures in the county filled with swine in "lounging sheds" or feed cattle pens adjacent to corn and soy fields. At the end of the century, however, prevailing market conditions, new environmental regulations, and the trend toward agricultural specialization meant that it was rare to find a farmer who devoted attention and money to raising both grain and commercial livestock. Perhaps one of the most obvious examples of the changing times and specialization of agriculture are the massive blue Harvestore silage tanks that signal that the farm once raised livestock. Today, few DeKalb farmers raise any livestock, and virtually all the blue tanks now stand empty—on one old farmstead west of Sycamore, the enterprising owner cut windows and a door into the squat Harvestore tank, built a porch, installed a heater, and recycled the old tank into a new room.

The specialization of agriculture also meant that farmers became oriented even more to national and international markets. National concerns shared attention with local ones. Finally, farmers in the mass society period began to borrow money to a much greater degree to finance their increasingly costly operations. In many respects these financial transactions resembled more the practices of corporations, rather than the intimate relationship that once existed between farmer and local banker, and reflect the extent to which farming had transformed into a business.

At the end of the twentieth century, the nation, including the agricultural sector, was witnessing the beginning of a third social era, that of the information age. The speed of data transfer, the amounts of transmitted information, high-quality long-distance connections, the ability to access current information from nearly any location, the importance of electronic communication and commerce, and access to specialized data are just some important features of this fundamental revolution in American society. We have shifted into an industrial economy in which information is as important as, if not more important than, production— one needs only to look at the nation's balance-of-trade figures to see how much American innovation is delivered offshore to be converted by others into practical applications. Information is substituted for time, labor, and energy in the production of farm products. Between 1970 and 1980, nearly 90 percent of new jobs created in the country were described as information or service activities.[5] It is difficult to predict in what directions the new information era will take us, and it is likely that the linking of technologies through information improvements will lead to new capabilities, demands, and products. It is in this information-saturated environment that the Farm Bureau will need to act in response to a host of old and new interrelated social, technical, and economic issues confronting its members.[6]

In recent years, one of the most significant transformations of rural America has been a dramatic population increase after decades of population stability or decline. Between 1970 and 1980, for the first time since the 1820 census, the population of what were considered to be rural counties increased faster than the urban populations, and the trend has continued to accelerate. Yet, as populations in counties at the margins of metropolitan regions grow, the number of farmers continues to diminish as a proportion of the total population. The rapid growth rate of DeKalb County at the turn of the twenty-first century is hardly unusual in the Farm Belt. The population of farmers decreased nearly 40 percent during that same decade.[7] This "gentrification" of rural America has already had a profound effect on country folk, and the pace of change is likely to accelerate in the future, especially in communities within the economic orbit of major cities.

One consequence of the inflow of newcomers is that rural nonfarm issues become increasingly important. County roads, for example, need to be planned more for residential commuting and commercial traffic patterns than for farming ones. Housing and industrial development and the expansion of city and town boundaries is already a perennial debate for DeKalb County residents. Moreover, as the percentage of farmers declines in proportion to the total population, nonfarm interests will have the political power to influence how agricultural resources are used. New zoning laws, for example, may curtail farm operations in favor of residential landholders—as land-use patterns in the county change, yesterday's successful swine operation is tomorrow's public nuisance. Zoning regulations designed to curb urban sprawl by preserving farms and creating greenbelts between cities may seriously impair the marketability of farmland. The exercise of political power at the expense of agricultural interests is certain to elevate tensions between farmers and newcomers. One important function of the Farm Bureau will be to continue to act as an advocate for its agricultural members. Part of that representation will likely require it to act as a broker between competing interests in the county. The real difficulty for the Bureau may be in helping farmers balance their expectations with contemporary demographic conditions.

Another important consequence of new population growth is that communities like DeKalb County, which historically rely on agriculture as their primary source of wealth, are likely to find it difficult to pay for public services. Farming land will transfer to nonfarming holders, who will utilize the land for alternative uses that change its tax obligations.

Rural towns experiencing population increases are required to provide for new or expanded services, such as sanitation, schools, transportation, and public safety while simultaneously confronting a diminishing nonland tax base. Moreover, in many places the land-use change is from agriculture to residential, boosting demand for expensive infrastructure improvements while excluding commercial development and tax revenue. Ironically, rising demand for new services is occurring at the same time that costs for those services are escalating. In addition, new government regulations, many of which are unfunded mandates, will place even more cost burdens on taxpayers. The pressure to raise new public revenues from agriculture lands and products has been a political problem for farmers for generations, and it is likely to increase even more in the future. The Farm Bureau already has tremendous historical experience in the areas of support for community services and tax policies that protect rural real estate from what farmers consider unreasonable taxation. In the future, Bureau members will need to be even more assertive in voicing their collective concerns on these issues through the Farm Bureau if they hope to influence the political landscape of the future.

There are also important demographic changes that are shaping the Farm Bureau from within. Membership has remained relatively stable in recent years, but, like DeKalb County as a whole, the percentage of farmer members is gradually declining. Bureau membership is increasingly dominated by "Class A" Associate and "Class A+" Associate Plus members who share in many Bureau benefits, especially insurance and social programming, but have few governing privileges within the organization—a large percentage of Bureau members belong simply to purchase insurance and have little interest in Bureau affairs at all. The number of "Class M" members, those who are landowners and whose occupations are principally agricultural and are therefore entitled to vote on important Bureau matters, has declined in recent decades, concentrating institutional power in fewer hands and accentuating the stratified membership structure. That differential structure can result in internal divisions among members, with agricultural members supporting issues that help them, while nonfarm members may favor alternatives. Moreover, as DeKalb County becomes more urban, there are alternative service providers and social opportunities for residents, making Farm Bureau offerings simply part of a growing menu of social and financial alternatives. As fewer young people join the Farm Bureau, the average age of the Bureau membership is rising, and there are fewer and fewer lifelong senior members each year. This graying of the membership drains valuable experience and energy from the organization. It also will influence what offerings and benefits the Bureau provides in the future. Even agricultural newcomers to DeKalb County have not joined the Bureau in any significant numbers. Farmers who have sold their land in counties that ring greater Chicago and bought land in DeKalb County sometimes have little personal attachment to the DeKalb County Farm Bureau, eroding farmer-member representation in the organization still further. In the future, the DeKalb County Farm Bureau will likely need to work to sustain active membership while also providing incentives for new members if it is to remain vital into the coming decades.

Technical changes will continue to influence farming practices in the future. The acceleration of agricultural technology is not new. Innovation is a hallmark of twentieth-century agriculture, and ever since William Eckhardt's first day as farm advisor, the Bureau has been helping farmers master new techniques. The pace of change, however, has quickened in the last thirty years, and, in the future, it is likely that such changes will have even more pronounced consequences for farmers. "No-till" cultivation has become more common. Drought-resistant crop varieties are widely available. Carefully bred seed produces grains that have particular, and predictable, commercial properties. Farm equipment, much of it operated partially or entirely with computers, is larger, faster, and more efficient. New chemical formulations and more effective pesticides all present unique challenges. Genetically modified crops and animals are unique technologies that promise a new revolution in agriculture on par with the development of the plow, domestication of animals, and selective crop breeding. Genetic engineering in particular is already the subject of heated political debate. The concept of shifting genes from one species of plant to another, or between animals, or even between animals and plants, poses serious ethical debates among policy makers, environmentalists, and farmers. Nevertheless, genetically modified crops are a fact of life, and it will be the responsibility of future farmers to address the complex issues, real and imagined, that rise from them. In the new information era in which nearly everyone has access to the same information with a few keystrokes of a computer, the Farm Bureau will need to assist farmers less with acquiring information on the new methods and perhaps more on interpreting it.[8]

In addition, innovation on the farm will continue to reduce the unit costs of farming and expand production. Such production, coupled with inelastic demand and fluctuating markets, puts downward pressure on farm incomes that will continue. This is hardly a new phenomenon, and, as was the case in earlier decades, farm families will need to turn to nonfarm income sources to maintain their living standards. Farming for most in DeKalb County has already become a part-time career, not a way of life. Already, it is nearly impossible to find a DeKalb farm family that does not rely on nonfarm wages for a significant portion of the household income. Job-training programs, flexible employment, and other workplace innovations are likely to be a feature of rural society in the future. County institutions and organization, like the Farm Bureau, will need to work aggressively to attract new job opportunities to rural areas and county towns. On the other hand, in many cases, changes in rural job opportunities may be driven not by the need to supplement farm income, but rather by the very success of the farm itself. Wealthy operators or agricultural corporations, for example, may find it wise to diversify their incomes by investing in nonfarm enterprises. Farm family members may see jobs, especially professional ones, as complementary, not alternatives, to their farming base. In either case, the Farm Bureau will have to continue its long-term commitment to education, both for the sake of its members and the broader community. Moreover, it must continue to look to its roots in the regional commercial sector and work to attract high-wage jobs to the community that will benefit everyone.

The influx of nonfarmers, coupled with the need for farmers to become ever more cost efficient, influences how farmers hold on to their land. The increasing number of forces that determine agricultural uses—whether leases, zoning, or regulation—makes the price of farmland volatile. Because land equity has always been the basis of farm borrowing, such volatility can have disastrous consequences for farmers and supporting industries. One solution is to sell land to raise capital, and, especially near metropolitan centers, the pressure to sell the homestead and buy new land further away, or simply retire, is great. Selling a parcel of land, of course, is a one-time solution to an inherent and institutional problem of equity and debt that limits future land-use choices. Farmers who desire larger parcels of land to maximize their operations increasingly lease fields rather than buy them. Often, much of that land available for rent is held by people or corporations who take a more active role in managing their land than did lessors in earlier times. Farming landowners may rent their land and equipment to others in ways that were even less common in earlier decades. It is not unusual, for example, for a DeKalb livestock operator to lease his facilities to someone else and accept a wage for tending another's animals on his own land.

In addition, federal and state governments have also reduced the risks of owning and transferring farmland in other ways. Crop programs, insurance programs, direct food purchases, and market development here and abroad have all benefited the agricultural sector. The Farm Bureau may find it helpful for area farmers to continue its financial management programming, and even expand it to include education about the rapidly evolving capital markets.

Another concern confronting DeKalb county agricultural residents is the shift in the nature of farm markets. Long-held assumptions about market competition are increasingly out of step with reality. It is a trend that has been under way throughout the life of the Farm Bureau—McNary-Haugen proposals, the New Deal, post–World War II federal programs, and the like were all challenges to traditional American faith in agricultural entrepreneurship. Each governmental initiative, however, was at least cloaked in the rhetoric of capitalism that envisioned family farmers as the core of the agricultural sector. Today farmers have even less control over their markets than ever. More than ever before, commodity and processing firms, not producers, make decisions about what to raise in the field. Continued consolidation of farming into the hands of large corporations has changed the nature of price competition. New brokers and bargaining institutions are likely to negotiate details of exchange with farmers who have little commercial leverage. Indeed, it is that very problem—the inability to master the evolving art of matching product with willing buyers in the information age—that contributed to the demise of DCX. Traditional models of farm competition are likely to linger as a skeleton for the new farm economy, but even now DeKalb County farmers are experiencing market trends that would have baffled farmers a generation or two ago. Historically, the Farm Bureau faced a variety of marketing changes and, on occasion, exerted its political and economic influence to shape those changes and buffer area farmers from their worst consequences. In the coming years, it is likely that such aggressive action will be even more necessary to help shield its members from the worst consequences of such rapid changes.

Environmentalism shaped farming in the past half century and is likely to impact farming in the future. Farmers and ranchers, both family and corporate, manage two-thirds of America's land and are the primary stewards of much of the nation's soil, water, and air. Farmers, more than many people, have long-term incentives to maintain, protect, and enhance the resources upon which they depend for their livelihood— they have a powerful motivation to do the right thing. In fact, it was the desire to conserve the soil fertility and sustain productivity of DeKalb County agriculture that was at the heart of the original Soil Improvement Association and has been a core topic of Farm Bureau programming throughout the decades. American farmers have certainly been responsible for much environmental damage and often resisted environmental initiatives to solve those problems. The American Farm Bureau Federation, for example, was a regular critic of public efforts to curtail environmentally hazardous farming techniques throughout the 1960s and beyond. The farming sector, however, has also achieved many notable successes. In general, farmers use fewer expensive chemicals and pesticides now than they did in previous decades—no one today believes the inflated promises of the 1950s, which predicted a chemical solution to nearly every farm problem. Today, chemicals are better fitted to the desired outcomes and pose fewer sustained environmental pollution issues. The coupling of selective breeding and genetic engineering with sophisticated biochemistry has produced plants that are immune to the effects of certain chemicals. Farmers are now able to tailor their herbicide applications more accurately and consume less fuel for repeated spraying or mechanical weed control. One such genetic modification important for DeKalb County was the development of "Roundup Ready" soybeans. These soy plants, pioneered by DeKalb Genetics, the successor to the DeKalb County Agricultural Association, are unaffected by the application of an herbicide designed to clear weeds from the bean rows. No-till cultivation reduces chemical runoff and soil erosion.[9]

Environmental self-interest is also reinforced by government actions, especially since the 1960s. For example, federal programs offer incentives to farmers to preserve wetlands—despite the shrill cry from environmentalists, wetland destruction per year between 1992 and 1997 was less than one-twentieth what it averaged annually between 1954 and 1974. Federal subsidies assisted research and development and encouraged purchases of advanced technologies. Private conservation groups, like the Nature Conservancy, work in partnership with farmers to set aside special lands. The Farm Bureau, which has a long conservation history in DeKalb, including support for forest preserves and wildlife, should continue its environmental-quality programming to balance land demand with resource conservation.[10]

At the same time, however, environmental action comes with costs. New technologies are expensive, although they often result in higher yields. During the past thirty years, federal and state involvement in the environment changed from offering farmers advice and supporting voluntary conservation to more coercive action. Increasingly, conservation practices are enforced by criminal and civil penalties. Especially since the 1980s, farmers face a new arena of risk in which their traditional activities—clearing land, plowing, draining fields, controlling pests, fencing property, fertilizing, and harvesting, storing, and transporting crops—expose them to sanctions that could bankrupt a farming corporation, much less a family farmer. Farmers must also be careful not to become ensnared in questionable environmental claims. In the farm belt, for example, there is today considerable support for converting crops to ethanol for use as an alternative fuel supply and to reduce pollution. In reality, however, evidence suggests that environmental costs are greater than advertised and may outweigh ethanol's claimed benefits.[11] In their days, farm advisors like Eckhardt and Roberts cautioned farmers against quacks, patent medicines, questionable soil improvement chemicals, and dubious conservation methods. Today, the Farm Bureau must be even more vigilant in investigating environmental claims and provide a respected moderating voice in highlighting both the benefits and consequences of environmental regulation.

This is only a partial listing of the issues that the Farm Bureau must address in the coming years. There certainly will be many problems that cannot be foreseen, but these projections are rooted in history and are already in process. Likewise, the Farm Bureau has experience in helping its members confront the problems of modernity that it can draw upon as a tool for resolving future challenges. In one area, however, there is little precedent in DeKalb County Farm Bureau history. All the changes, known and unknown, can be addressed only by a Farm Bureau that remains confident of its abilities. It was confidence, after all, that was the DeKalb County Farm Bureau's most consistent quality throughout the twentieth century. The original Soil Improvers always maintained a faith that their ideas were correct and their goals achievable. The Farm Bureau advisors believed that they held the keys to the long-term prosperity of the county's fields and herds. No economic climate was so bad that smart farmers who followed the Bureau's recommendations could not survive, and even prosper. Collective-marketing associations could ease the consequences of high production. The Bureau created and directed a variety of business concerns. Wartime production quotas were never impossible to fill. Federal farm policy could be influenced by a handful of DeKalb farmers who were led by a man who believed taking on Washington was the surest way of leaving a farming future to his children. Local farmers with enough determination, cash, and organizational support could start an international grain-marketing firm to compete with the multinational agricultural corporations. Regular and lavish community charity was essential. These are the actions of an organization whose leaders and members are utterly confident that their cause is just, their ideals can be achieved, and hard work will make the future better than today.

Such supreme confidence is partially rooted in historical notions about American agriculture. In traditional American thinking, farmers were the elite of society. Farmers were good citizens. The nation should be home to many farmers. Farming was a way of life and a family enterprise. Landownership is preferable to leasehold. Production should increase. Anyone who wants a farm should have one.[12] Farming was elevated to a spiritual, redemptive level. "Those who labor in the earth are the chosen people of God," wrote Jefferson, "whose breasts He has made His particular deposit for substantial and genuine virtue. It is the focus in which he keeps alive that sacred fire, which otherwise might escape from the face of the earth. Corruption of morals in the mass cultivators is a phenomenon of which no age nor nation has furnished an example. . . . While we have land to labor then; let us never wish to see our citizens occupied at a work bench, or twirling a distaff. . . . The mobs of great cities add just so much to the support of pure government, as sores do to the strength of the human body."[13]

Throughout most of the DeKalb County Farm Bureau's existence, many people in the county would have agreed with Jefferson. Probably some believe it now. Whether they favored expanded government involvement in American agriculture or preferred private market capitalism, they all viewed farming as the core of the nation and an occupation that was worth preserving as nearly as possible in the eighteenth-century image. In light of their fundamental faith in the nobility of agriculture, it is easy to understand how DeKalb County farmers believed that they could shape their future through collective action. In recent years, the American social understanding of agriculture changed. The nature of farming, and the very identity of farmers as an important part of American society, eroded. In the face of these changing assumptions, the DeKalb County Farm Bureau will need to retain confidence in itself and its mission if it is to remain a viable agricultural organization.

The Jeffersonian ideal of family farms and the moral superiority of rural folk is a central feature of American culture's beliefs about farming, but however important that was, it alone does not explain the DeKalb County Farm Bureau's decades of success. There is no question that the Bureau also benefited from its financial position. DeKalb County has always been one of the wealthiest agricultural counties in the state. The early relationship between the Farm Bureau and the DeKalb County Agricultural Association, especially the initial Bureau investment in Association stock and the conservative financial decisions by the Bureau leadership and the Rural Improvement Society, empowered the DeKalb County Farm Bureau to accomplish far more than most other bureaus across the nation. The DeKalb County Farm Bureau was one of the best endowed in the country, a fact that enabled it to fund its programs generously. Money never guarantees success—in fact, the organization's deep pockets and overconfidence on occasion led to some costly mistakes—but it did provide the Bureau leadership with enough security to take the institution into new areas while sustaining its basic mission and programming.

Bureau members also gained confidence from the structure of the organization itself. Historically, the Bureau was highly democratic, with thirty-eight directors who represented all nineteen county townships. Voting members took their voting responsibility seriously. "All politics are local," goes the old saying, and the election for Farm Bureau board representatives was as important, if not more important, for many members as any governmental election. The intimacy of the Bureau and of the county as a whole during the Bureau's early decades, the generally open nature of the organization, regular membership meetings, published reports, and the monthly *DeKalb County Farmer* meant that members were informed about Bureau proposals and actions. For decades, local television, radio, and newspapers reported Bureau activities. Some of the earliest supporters of the Soil Improvement Association, after all, were the area's newspapermen. There was also considerable opportunity for individual members to voice their concerns directly to the leadership.

For their part, the men who sat on the Bureau's board of directors were deeply dedicated to the organization. For them, farming and livestock were a way of life as much as a business, and the Bureau was simply another inherent feature of that farming reality. There was also a hierarchy of power within the board by which new members served a sort of apprenticeship, learning the Bureau's operations and how to govern the organization before taking greater responsibility. Finally, there was great continuity within the leadership. Bureau politics could occasionally be intense enough to breed years-long animosities, and even cause members to leave the Bureau, but it was seldom openly hostile. Few directors were forced out of office by their constituents, election challenges to incumbent directors were rare, and upset elections were rarer still, all of which signal that most members retained confidence in their

Bureau representatives. Most directors served long terms, sometimes un-interrupted for decades. Many who served were second-generation rep-resentatives. Some members complained from time to time that the board was autocratic, but such longevity also meant that the leaders were acquainted with each other's strengths and weaknesses, as well as with the formal and informal power structure of the Bureau. The other area of leadership support came from the farm advisors. During the early days of the Soil Improvement Association, the relationship between the board and the advisor was close. The Association, and later the Bureau, hired the advisor and paid the salary. It had been that way ever since the founding Soil Improvers decided to remain an independent farm organization free from government ties. Men such as Eckhardt and Roberts, however, were more than employees. Although the advisors could not vote on Bureau board policies, they often exercised considerable influence. They offered suggestions and played a role in directing Association and Bureau policies for decades. That close relationship continued until the "Great Di-vorce" in the early 1950s. Even then, however, the alliance persisted and the board members continued to rely on the formal and informal counsel from the advisors. Some advisors also stayed in DeKalb for years, further ce-menting a solid core of Bureau leadership. Such responsive membership and stable administration resulted in a Bureau philosophy and program that usually reflected the wishes of a majority of its members. It is little wonder, then, that the Bureau found it easy to undertake such a wide vari-ety of responsibilities throughout the decades.

The DeKalb County Farm Bureau drew confidence from its own ac-complishments. From the very beginning of the Soil Improvement Asso-ciation, the Bureau has enjoyed an enviable record of success in a vari-ety of areas. The Soil Improvement Association became the organizational model for later associations and bureaus in counties across the nation. Its farm advisors were always among the most tal-ented. Its adult and youth education programs were imaginative and carefully revised to address the needs of DeKalb farmers and their fami-lies. Most of its business ventures flourished, and when they did not, the Bureau cut its losses with little regret. Regular membership levels fluctu-ated, but they never reached a low point that threatened the existence of the Bureau. DeKalb 4-H children regularly competed at state and na-tional levels. Even its sports teams won. It has been repeatedly recog-nized as the best farm bureau in Illinois. For decades, it was the unques-tioned collective voice for area farmers on a broad variety of local, state, and national issues.

Perhaps most important of all, members of the Farm Bureau historically had a deep affection for the organization that was rooted in the personal relationships members had with it. Throughout its history, the DeKalb County Farm Bureau was always more than simply an agricultural assistance office for its members. Farm advisors after Eckhardt were committed to the welfare of the entire farm family. The Bureau regularly supported the local business sector directly and informally. The Bureau's educational focus bolstered local school reforms and the expansion of higher education in the region. Civic charities benefited from Bureau patronage. In the early years, many residents had a direct hand in the creation of the Soil Improvement Association. Many more could attest to the improvements Bureau membership had made in nearly every facet of their lives—for many DeKalb families, the Bureau and the programs it sponsored influenced the very structure of their lives from childhood through retirement. The DeKalb County Farm Bureau evolved into a community within a community. Members joined and quit throughout its history, but there was always a nucleus of members who considered their Bureau to be an extended family in which siblings disagree and occasionally bicker but just as often agree and always come together at times of crisis to defend their beloved Bureau.

Which ultimately brings us back to the black DeKalb soil that is the heart of the DeKalb County Farm Bureau story. For millennia, all kinds of people prospered on the land. In more recent times, Indians and whites fought and killed and died for it. Farmers prospered. Agricultural industries derived wealth from the surrounding farmland. More than a century and a half of agriculture on the fertile and gently rolling prairie fields created in generations of farmers an intense identity with the land that only people who depend on the soil for their lives can understand. That earth is what prompted the creation of the Soil Improvement Association in the first place. It is what sustained the prosperity of the region throughout the twentieth century. And it is the members' pride of place and drive to succeed on the land, their native soil, that will ultimately determine whether the Bureau can sustain itself in an uncertain future.

APPENDIX

Former DeKalb County Farm Bureau Officials

DeKalb County Farm Advisors

1912–1920	William G. Eckhardt
1920–1931	Thomas H. Roberts
1931–1936	Russel N. Rasmusen
1936–1942	Roy P. Johnson
1942–1944	Donald G. McAllister
1944–1949	W. Carroll Mummert
1949–1952	Ward H. Cross
1952–1954	H. Clifford Heaton
1954–1970	Elroy E. (Al) Golden *(1954, Farm Bureau and Extension separated)*
1970–1976	John I. Goodrich
1976–1980	Kenneth R. Bolen, *Extension agricultural advisor*
1981–1983	Lyle Paul, *Extension agricultural advisor*
1983–1992	David Whitson, *Extension agricultural advisor*
1992	*Farm advisor's position changed to a regional Extension position.*

DeKalb County Home Advisers

1937–1938	Charlotte Biester
1938–1939	Venus Johnson Evans
1939–1943	Helen C. Johnson Ogilvie
1943–1952	Bernice Hayes Engleking
1952–1952	Jean Neese Pickett
1952–1956	Esther Siemen McKellar *(1954, Farm Bureau and Extension separated)*
1956–1959	Delores Parrott Cantlon
1959–1961	Jessie M. Nixon
1962–1985	Deloris J. Ellis, *Extension home economics advisor*
1985–1992	JoAnn Skabo, *Extension home economics advisor*
1992	*County home advisor's position changed to a regional Extension position.*

Farm Bureau Presidents

1912–1923	Dillon S. Brown
1923–1924	Henry H. Parke
1924–1928	Henry J. White
1928–1931	Aaron J. Plapp
1931–1933	L. D. Sears
1933–1946	E. E. Houghtby
1946–1950	George Bark
1950–1956	Lloyd B. Waldee
1956–1974	Howard Mullins
1974–1977	Donald Chaplin
1977–1986	Allan Aves
1986–1997	Ken Barshinger
1997–2002	Vince Faivre
2002–present	Paul Rasmussen, Jr.

County Organization Director—Secretary of Organization— Executive Secretary—Manager

1941–1952	Lee Mosher, *county organization director and livestock spec* *(1952 became only livestock specialist)*
1952–1963	Henry M. Carlson, *county organization director*
1963–1964	J. Leslie Logan, *secretary of organization*
1964–1967	Joseph E. Hodges, *executive secretary*
1968–1969	Ronald Jensen, *executive secretary*
1969–1976	Melvin Hayenga, *executive secretary*
1976–1978	Robert H. Muehling, *executive secretary*
1979–present	Douglas M. Dashner, *manager*

NOTES

Introduction

1. To avoid confusion, it is important to note here that DeKalb County's largest city is also named DeKalb.

2. "Kishwaukee" was Algonquian for "sycamore," trees that were numerous in the area at the time of white settlement. J. G. Mosier, H. W. Stewart, E. E. DeTurk, and H. J. Snider, *DeKalb County Soils: Soil Report No. 23* (Urbana: University of Illinois Agricultural Experiment Station, 1922), p. 8; Leonard J. Kouba, *Resource Use and Conservation in DeKalb County* (DeKalb: DeKalb County Conservation Education Committee Print, 1963), pp. 3–20; Robert M. Grogan, "Present State of Knowledge Regarding the Pre-Cambrian Crystalices of Illinois," *Short Papers on Geological Subjects* 157 (1950): 97–98; William S. Lawrence and Associates, Inc., *Comprehensive Plan for DeKalb County, Illinois*. U.S Department of Housing and Urban Development project number P-268 (DeKalb, Ill.: DeKalb County Planning Commission, 1971), pp. 1–32.

3. Kouba, pp. 15–18

4. Several prairie parks in DeKalb and surrounding counties preserve fragments of original tall grass prairie for modern visitors. As impressive as they are, however, none is large enough to convey anything but a fleeting impression of what was once an ecosystem that encompassed hundreds of square miles of northern Illinois.

5. U.S. Department of Agriculture, *Illinois County Agricultural Statistics, DeKalb County* Bulletin C-11 (Springfield: Illinois Department of Agriculture, 1965), p. 4; Kouba, pp. 32–60; Mosier et al., pp. 3–4.

6. Teosinte still grows wild in remote areas of Mexico, and genetic evidence established that modern corn, *Zea mays*, is a direct descendant. More recent and sophisticated genetic profiling suggests that modern corn may be a cross between teosinte and gamagrass, or *Tripsacum*. The archeological record indicates that maize evolved very rapidly, perhaps over only a century, giving weight to this theory. Rather than the slow, deliberate breeding program leading to maize, a fertile cross between teosinte and gamagrass could have yielded early versions of maize within a relatively short time. Humans would have selected plants from the random crosses that had the best seeds and reproduction rates. See, for example, Mary Eubanks, "An Interdisciplinary Perspective on the Origin of Maize," *Latin American Antiquity* 12:1 (2001): 91–98, and "The Mysterious Origin of Maize," *Economic Botany* 55:4 (2001): 492–514.

7. The Hopewell civilization appeared around 300 BCE and lasted for about seven hundred years. Hopewell culture may have emerged from an earlier indigenous one, the Adena culture, or it may have come from Mexico up the interior rivers, especially the Mississippi. At its peak, Hopewell culture spread across North America in a broad sweep from Florida to the Great Lakes region of southern Canada, and from the lands west of the Appalachian Mountains to eastern Nebraska. There are several theories suggesting the causes for the demise of the Hopewellians, including drought, disease, and warfare, but there is little conclusive evidence supporting any theory. See, generally, Thorne Deuel, "The Hopewellian Community," *Hopewellian Communities in*

Illinois, Scientific Papers. Vol. 5 (Springfield: State of Illinois, 1952); Illinois Archeological Survey, *Illinois Archeology, Bulletin 1* (Urbana: University of Illinois, 1973).

8. Dorothy Anne Dondore, *The Prairie and the Making of Middle America: Four Centuries of Description* (Cedar Rapids, Iowa: Torch Press, 1926), pp. 37–40.

9. *DeKalb Daily Chronicle*, September 10, 1934. Unfortunately, Captain Webb's first name is not known.

10. Two sections of land, about 1,280 acres, at Paw Paw Grove, however, were reserved for Chief Shabbona and his family in gratitude for his friendship and assistance to the early white settlers during the Black Hawk War. Those lands were confirmed to Shabbona in 1832. By 1840 white gratitude had worn thin, however, and the General Land Office commissioners decided Shabbona had forfeited his land title, and the property was auctioned to white settlers for $1.25 per acre, considerably less than it was worth. See Harriet Wilson Davy, *From Oxen to Jets: A History of DeKalb County, 1835–1963* (Dixon, Ill.: Rogers Printing, 1963), pp. 207–9.

11. The earliest permanent white settlers arrived in DeKalb County from many parts of the country. William Sebree, for example, was a Virginian. In general, however, DeKalb County was settled primarily by native-born whites originally from New England, New York, and other Northern states. Those settlers retained their Yankee attitudes and Protestant moral earnestness. During the 1840s and 1850s, DeKalb County, and especially the towns of Sycamore, Mayfield, and South Grove, were sympathetic to the abolitionist cause, and many local farms were "stations" on the Underground Railroad. DeKalb County voted for Abraham Lincoln in the 1860 election and was fiercely pro-Union during the Civil War. See Davy, pp. 3–22.

12. Davy, pp. 3–12; Lewis M. Gross, *Past and Present of DeKalb County, Illinois* (Chicago: Pioneer Publishing, 1907), pp. 29–31; *Biographical Record of DeKalb County, Illinois* (Chicago: S. J. Clarke, 1898), pp. 401–3, 425–27.

13. Morris Birkbeck, *Letters from Illinois* (New York: Da Capo Press, 1970), p. xi.

14. Dondore, pp. 206–7, 213; Davy, p. 3.

15. Betty Flanders Thomson, *The Shaping of America's Heartland* (Boston: Houghton Mifflin, 1977), pp. 164–77; Kouba, p. 61; James E. Myers, "Our Prairie Past," *Chicago Tribune Magazine* (August 21, 1977), p. 45; Frederick Gerhard, *Illinois as It Is* (Chicago: Keen and Lee, 1857), p. 121; *DeKalb Daily Chronicle*, September 10, 1934.

16. When Illinois was admitted to the Union in 1818, the land that is now DeKalb County was part of St. Clair County. Between 1818 and 1837, the DeKalb area at one time or another was part of Madison, Edwards, Crawford, Clarke, Pike, Fulton, Putnam, LaSalle, and Kane counties. DeKalb County is named for Baron Johann DeKalb, a German officer who died at Camden, South Carolina, in 1780, while fighting in the Continental Army. See Davy, pp. 29–31.

17. Jesse C. Kellogg and Henry Lansom Boies, *DeKalb County History, 1871* (Genoa, Ill.: Thompson and Everts, 1871), p. 6.

18. Ibid., p. 7.

19. The National Park Service named the John Deere homestead in Grand Detour, Illinois, where Deere invented his famous plow, as a National Historic Landmark. The American Society of Agricultural Engineers designated it as a Historic Landmark of Agricultural Engineering. Today, visitors can view the archeological remains of the blacksmith shop and the original Deere home and grounds.

20. For example, in 1845, DeKalb County assessed $375 in taxes, $362 of which was paid using jurors' certificates or other county obligations. Henry Lansom Boies, *History of DeKalb County, Illinois* (Chicago: O. P. Bassett, 1868; reprint, Evansville, Ind.; Unigraphic, 1973), pp. 399–408.

21. Boies; U.S. Department of Commerce, *The Seventh Census of the United States, 1850* (Washington, D.C.: Robert Armstrong, 1853); Bertha G. Bradt, "Comments on

Economic Conditions," (Box 1, Bertha G. Bradt Collection, Regional History Center Archives, Northern Illinois University, DeKalb); U.S. Department of Commerce, *Population of the United States in 1860; Compiled from Original Returns of the Eighth Census* (Washington, D.C.: Government Printing Office, 1864); Journal from 1854 (Box 2, Hall Family Collection Regional History Center Archives, Northern Illinois University, DeKalb); Journal from 1855 (Box 2, Hall Family Collection); U.S. Bureau of the Census, *Historical Statistics of the United States, Colonial Times to 1970, Bicentennial Edition.* Part 1 (Washington, D.C.: Government Printing Office, 1975); Lewis Severson, "Some Phases of the History of the Illinois Central Railroad Company since 1870," Ph.D. dissertation, University of Chicago, 1930. Although the railroad boosted the region's prosperity, it alone could not insulate the region from downswings in the economy that occurred as a consequence of the Panic of 1857, a general crop failure in 1858, and depressed agricultural profits in 1859.

22. James E. Boyle, Ron G. Klein, and Stephen Bigolin, "DeKalb County Courthouse," historical brochure (n.d.), author's personal collection.

23. The following works provide a good introduction to DeKalb County history: *Biographical Record of DeKalb County, Illinois;* Boies; Davy; Kellogg and Boies; *DeKalb Chronicle Illustrated Souvenir Edition, 1899* (reprint, Evansville, Ind.: Unigraphic, 1981); Gross; Charles W. Marsh, *Recollections, 1837–1910* (Chicago: Farm Implement News Company, 1910); Arthur C. Page, *The Illinois Farmer Book of DeKalb County, 1925* (Chicago: Orange Judd Farmer, 1925); *Portrait and Biographical Album, DeKalb County, Illinois* (Chicago: Chapman Brothers, 1885); *True Republican and Sycamore, Illinois Illustrated Prospectus, 1906* (reprint, DeKalb, Ill.: DeKalb County Historical Society, 1980); *Voters and Tax-Payers of DeKalb County, Illinois* (1876; reprint, Evansville, Ind.: Unigraphic, 1977).

1—A New Era

1. *DeKalb Daily Chronicle,* July 1, 8, 1911; U.S. Department of Commerce, Bureau of Census, *Thirteenth Census of the United States, 1910: Abstract of the Census with Supplement for Illinois* (Washington, D.C.: Government Printing Office, 1913), pp. 573, 578–79; Robert H. Wiebe, *The Search for Order* (New York: Hill and Wang, 1967), pp. 1–10.

2. *DeKalb Daily Chronicle,* July 10, 1911. In June 1858, C. W. and W. W. Marsh, two DeKalb County farmers, began constructing wheat harvesters that were a significant improvement on the McCormick reaper. The Marsh harvester did everything the McCormick reaper did, plus made binding more efficient. It changed the process from merely reaping to harvesting. The machine was so simple to operate that "a boy or girl could drive and bind from eight to ten acres a day," effectively cutting labor costs in half. By the late 1870s, the Marsh brothers adapted an automatic twine binder to their machine that replaced hand binding, making it possible for one man to do the work of thirty. Charles W. Marsh, *Recollections, 1837–1910* (Chicago: Farm Implement News Company, 1910).

3. A tractor rig costing $2,000 in 1909 was a significant cash investment. According to the 1910 U.S. census, there were 251,872 farms in Illinois, with a value of implements and machinery totaling $73,724,074—the average Illinois farm had just $293 invested in implements and machinery, which they probably acquired over several years. Moreover, credit terms for implements were usually different than for land, and often less flexible and more costly. The average value of all Illinois farm acres in 1910 was about $95 per acre, and about $50 more per acre in DeKalb County. The national average for farm labor in 1910 was about $27 per month (without board). U.S. Department of Commerce, *Statistical Abstract of the United States, 1920* (Washington, D.C.: Government Printing Office, 1921).

4. *Sycamore True Republican,* September 29, 1909. The first tractor factory was built by Charles W. Hart and Charles H. Parr in Charles City, Iowa, in 1900. The early machines were bulky, weighing up to ten tons. For a few years, the machines were called "gasoline traction engines." In 1906, sales manager W. H. Williams rebelled against the awkward name, and was the first to refer to the new machines as simply "tractors," a name that has been used ever since. By 1910, thirty-one different manufacturers had built 2,000 tractors. By 1940, tractors had become so efficient that one man with a tractor and various plows, disks, harrows, planters, and cultivators could accomplish nearly three times the work of one man and five horses with less energy and labor cost. John Strohm, "Farm Power: From Muscle to Motor," *Prairie Farmer* 113:1 (January 11, 1941): 30. See generally, Robert C. Williams, *Fordson, Farmall, and Poppin' Johnny: A History of the Farm Tractor and Its Impact on America* (Urbana: University of Illinois Press, 1987).

5. Harriet Wilson Davy, *From Oxen to Jets: A History of DeKalb County, 1835–1963* (Dixon, Ill.: Rogers Printing, 1963), p. 101; *DeKalb Daily Chronicle,* September 10, 1934.

6. "Parity price" refers to the price of a farm commodity that would make the purchasing power of a unit of that commodity equal to its purchasing power at a given time in the past. Most parity calculations use the years 1909–1913 as the baseline for parity calculations, though the so-called parity period is 1910–1914. The Agricultural Adjustment Act of 1933 used the period July 1909–July 1914. Murray R. Benedict, *Farm Policies of the United States, 1790–1950* (New York: Octagon Books, 1966), pp. 114–15. See, generally, Joseph Davis, "An Evaluation of the Present Economic Position of Agriculture by Regions and in General," *Journal of Farm Economics* 15:2 (April 1933): 247–54.

7. U.S. Congress, *Report of the Country Life Commission,* Senate Document 705, 60th Cong., 2d Sess., 1909, p. 21.

8. Benedict, pp. 112–37; William L. Bowers, *The Country Life Movement in America, 1900–1920* (Port Washington, N.Y.: Kennikat Press, 1974), pp. 3–24; Willard W. Cochrane, *The Development of American Agriculture: A Historical Analysis* (Minneapolis: University of Minnesota Press, 1993), pp. 99–121; Donald C. Horton, Harald C. Larsen, and Norman J. Wall, *Farm Mortgage Credit Facilities in the United States.* U.S. Department of Agriculture, *Department of Agriculture Miscellaneous Publication No. 478* (Washington, D.C.: Government Printing Office, 1942), p. 1.

9. Bowers, pp. 7–15.

10. Ibid.; Benedict, p. 116; David B. Danbom, *The Resisted Revolution: Urban America and the Industrialization of Agriculture, 1900–1930* (Ames: Iowa State University Press, 1979), pp. 18–22.

11. Wiebe, pp. 125–27.

12. Bowers, pp. 3–29; Wiebe, pp. 125–27; William L. O'Neill, *The Progressive Years: America Comes of Age* (New York: Dodd, Mead, 1975), pp. 6–9.

13. *DeKalb Daily Chronicle,* September 20, 1921. Hill was one of the most vocal supporters of scientific agriculture as part of a national resource conservation movement. He feared that if farmers continued to exploit the soil, population growth would eventually overcome farm capacity with disastrous results. In 1910 he published *Highways of Progress,* in which he elaborated his views. He was sharply critical of the land-grant colleges and the Department of Agriculture for their inefficient dissemination of agricultural information. To fulfill his goal of making education accessible to farmers, he established five-acre experimental and demonstration plots along his rail lines and supported the county agent system. Bowers, p. 19. See also the serial version of Hill's book in "Highways of Progress; What WE Must Do to be Fed," *World's Book* 19–20 (November 1909–May 1910).

14. Bowers, p. 23; George F. Wells, "The Rural Church," *Annals of the Academy of Political and Social Science* 40 (March 1912): 131–39.

15. Bowers, pp. 77–79; Mabel Carney, *Country Life and the Country School: A Study of the Agencies of Rural Progress and of the Social Relationship of the School and the Community* (Chicago: Row, Peterson, 1912), pp. 114–19; U.S. Congress, *Report of the Country Life Commission*, p. 38; Harold Parker, "Good Roads Movement," *Annals of the American Academy of Political and Social Science* 40 (March 1912): 51–57.

16. Bowers, pp. 3–29; Cochrane, p. 112.

17. Minutes of the DeKalb County Agricultural and Mechanical Society, March 9, 1861, DeKalb County Supervisor's Proceedings, DeKalb County Court House.

18. Lincoln David Kelsey and Cannon Chiles Hearne, *Cooperative Extension Work* (Ithaca, N.Y.: Comstock, 1955), p. 14.

19. Ibid., pp. 11–15. See, generally, Alfred C. True, *A History of Agricultural Extension Work in the United States from 1785 to 1923*. U.S. Department of Agriculture Miscellaneous Publication No. 15 (Washington, D.C.: Government Printing Office, 1928); and Maurice C. Burritt, *The County Agent and the Farm Bureau* (New York: Harcourt, Brace, 1922).

20. State of Illinois, *Laws of the State of Illinois* (Springfield: H. W. Rokker, 1893), pp. 18–19.

21. *Seventeenth Annual Report of the Illinois Farmers Institute: A Handbook of Agriculture* (Springfield: Illinois State Journal Company, State Journal, 1912).

22. The *DeKalb Daily Chronicle*, for example, sponsored a corn contest in 1912 to "stimulate interest" in the new DeKalb County Soil Improvement Association and "lend 'moral support' to the betterment of conditions of our farms." The paper offered $50 in gold for the best bushel of corn raised in DeKalb County that year by a *Chronicle* subscriber. "The *Chronicle*," the writer modestly continued, "makes this prize offer that it may have a small part in the work making this county the greatest patch of farm land in the universe." In addition to the corn contest, the editors of the *Chronicle* offered their subscribers a year of the paper, a wall map of DeKalb County, a three-year subscription to the *Prairie Farmer*, and *Frank Mann's Soil Book* all for $2.50. In his widely circulated book, Mann, a regular speaker at DeKalb County Farmers' Institutes, director of the State Farmers' Institutes, associate editor of the *Prairie Farmer*, and successful Illinois farmer, provided practical guidance and stressed that careful soil and crop management could double farm production every year. *DeKalb Daily Chronicle*, April 24, 1912; Frank I. Mann, *Frank Mann's Soil Book* (Chicago: Prairie Farmer, 1912).

23. Bowers, pp. 15–44.

24. Wayne D. Rasmussen, *Taking the University to the People: Seventy-Five Years of Cooperative Extension* (Ames: Iowa State University Press, 1989), pp. 34–36.

25. Knapp's message was recorded on a marker at the Porter farm at Terrell, Texas.

26. Danbom, p. 72.

27. Ibid.

28. See, generally, Danbom, pp. 69–74; Benedict, pp. 152–53.

29. Danbom, p. 72.

30. Ibid., pp. 72–74.

31. Harry C. Sanders, "The Legal Base, Scope, Functions, and General Objectives of Extension Work," in Harry C. Sanders, ed., *The Cooperative Extension Service* (Englewood Cliffs, N.J.: Prentice-Hall, 1966), pp. 28–29.

32. Danbom, pp. 72–74; Benedict, pp. 153–54.

33. Orville M. Kile, *The Farm Bureau through Three Decades* (Baltimore: Waverly Press, 1948), pp. 31–34, and *The Farm Bureau Movement* (New York: Macmillan, 1921), pp. 94–99. Julius Rosenwald (1862–1932) amassed a fortune as a partner and executive of Sears, Roebuck and Company and became one of the most respected philanthropists of his era. He was especially interested in all types of education and

is credited with contributing more than $65 million to support educational and humanitarian charities. See, generally, Edwin R. Embree and Julia Waxman, *Investment in People: The Story of the Julius Rosenwald Fund* (New York: Harper and Brothers, 1949); Alfred Q. Jarrette, *Julius Rosenwald: Son of a Jewish Immigrant, a Builder of Sears, Roebuck and Company, Benefactor of Mankind* (Greenville, S.C.: Southeastern University Press, 1979).

    34. Kile, *Farm Bureau Movement,* pp. 97–98.

2—The "Soil Improvers"

    1. *Sycamore Tribune,* December 31, 1903; October 18 and December 20, 1904; *Sycamore True Republican,* April 20 and November 23, 1910; *DeKalb Daily Chronicle,* November 11 and 28 and December 5 and 8, 1911. Stock Certificate "DeKalb County Agricultural and Mechanical Society," issued to Sanford Taylor, 1871; Minute Book, Afton Center Alliance, March–December, 1890; Autobiography of Henry Hall Parke (17 manuscript pages, n.d.), DeKalb County Farm Bureau Archive (hereinafter, DCFB Archive). It is possible that DeKalb and/or Sycamore hosted farmer institutes earlier than 1903. The 1911 farmers' institutes were held in Elva, Genoa, Kingston, Kirkland, Malta, Sandwich, Shabbona, Squaw Grove, Sycamore, and Waterman. DeKalb County agricultural reformers were typical agrarian Progressives, and a brief overview of speakers at the November kickoff luncheon for the December farmers' institute suggests that they were drawn from the community elites: President George Gurler, owner of a state-of-the-art dairy; Secretary Henry Hall Parke, University of Michigan–trained biologist, successful farmer from Genoa, and member of one of the county's pioneer families; Frank Greely, president of the Clinton Township Farmer Club; Aaron Plapp, president of the Malta Farmer Club; Dr. John W. Cook, president of the Northern Illinois State Normal School in DeKalb; Arthur Dodge, banker and farmer from Malta; Dillon Brown, banker and farmer from Genoa; W. W. Coultas, county superintendent of schools (Coultas's wife, Mila, was Henry Parke's sister); and Burr Smiley, state representative. Moreover, the county's most influential newspaper, the *DeKalb Daily Chronicle,* ran enthusiastic feature articles about the institutes and speakers.

    2. *DeKalb Daily Chronicle,* December 5, 8, 1911.

    3. Ibid., December 13, 1911.

    4. Ibid., November 11, 1911.

    5. The Sycamore Farmers' Club was originally incorporated as a not-for-profit corporation by the Illinois secretary of state on December 7, 1909, as the "DeKalb County Farmers' Club," with its corporate address listed as No. 14, Daniel Pierce Building Street, Sycamore, Illinois. Its stated objective was "[t]o promote more profitable and more permanent methods of agriculture." On March 9, 1911, the name was officially changed to the Sycamore Farmers' Club. Document of Incorporation, State of Illinois, Miscellaneous Record Corporation Book, DeKalb County Courthouse, Sycamore, Illinois, December 7, 1909; Sycamore Farmers' Club, *Minutes: 1909–March 18, 1942,* February 18, 1911, DCFB Archive.

    6. *DeKalb Daily Chronicle,* July 19, 1911.

    7. Ibid., September 20, 22, 27, 1911.

    8. Ibid., September 20, 22, 27, October 9, 1911.

    9. Ibid., October 17, 1911, February 3, 1912, June 30, 1922. The influential newspapermen of the movement were W. H. Ray, of the *Shabbona Express,* president; C. D. Schoonmaker, of the *Genoa Republican Journal,* secretary; H. W. Fay, of the *Shabbona Express,* secretary; Frank W. Greenaway, of the *DeKalb Chronicle,* executive committee; R. D. Chappell, of the *Hinkley Review,* executive committee; Frank Lowman, of the

Sandwich Free Press; H. W. Fay and E. O. Fay, of the *DeKalb Review;* W. H. Riley, of the *Malta Record;* Bailey Rosette, of the *DeKalb Advertiser;* Claude O. Pike, of the *Sycamore Tribune;* and E. I. Boies, of the *Sycamore True Republican.*

10. Sycamore Farmers' Club, "Program and Premium List, Second Annual Mid-Winter Exhibit and Farmers' Institute, Under the Auspices of the Sycamore Farmers' Club" Second Annual, February 8–12, 1912, DCFB Archive. *DeKalb Daily Chronicle,* July 19, 1911, November 28, 1911, October 9 and 17, 1911, and September 20, 1921. Among the founding members of the DeKalb County editorial association were President C. D. Shoonmaker, *Genoa Republican Journal;* Secretary H. W. Fay, *Shabbona Express;* F. W. Greenaway, *DeKalb Daily Chronicle;* R. D. Chappell, *Hinkley Review;* and J. B. Castle, a banker from Sandwich and member of the DeKalb County Bankers' Association. *DeKalb Daily Chronicle,* October 17, 1911.

11. *DeKalb Daily Chronicle,* December 27, 1911.

12. Ibid., January 3, 1912.

13. Ibid., January 4, 1912.

14. Ibid. Eckhardt was a protégé of the famous soil expert, Cyril G. Hopkins, professor of agronomy and chemistry. See for example: Cyril G. Hopkins, J. E. Readhimer and W. G. Eckhardt, "Thirty Years of Crop Rotations on the Common Prairie Soil of Illinois," *University of Illinois Agricultural Experiment Station Bulletin, No. 125* (Urbana, 1908). Eckhardt's first contact with DeKalb County was in 1907, when John W. Cook, President of the Northern Illinois State Normal School (later Northern Illinois University), invited Eckhardt to be a guest speaker at the annual farmers' institute. *DeKalb County Farmer* 28:5 (May 1940).

15. John L. Richardson, "Banker Brown," *Better Crops with Plant Food (The Pocket Book of Agriculture)* (January 1928), p. 24, DCFB Archive.

16. *DeKalb Daily Chronicle,* January 3, 4, 5, 1912.

17. Ibid., January 6, 1912. The executive committee was selected from among the members of the three professional organizations that formed the backbone of the farm advisor support. Henry H. Parke of Genoa, George Gurler of DeKalb, and F. W. Leifheit represented the DeKalb County Farmers' Institute; C. D. Schoonmaker of Genoa, R. D. Chappel of Hinkley, and Frank Greenaway of DeKalb represented the newspapermen; Fred Townsend of Sycamore, Dillon Brown of Genoa, and Samuel and Charles Bradt of DeKalb represented the bankers and local business interests.

18. Ibid., January 6, 1912.

19. John J. Lacey, *Farm Bureau in Illinois: History of Illinois Farm Bureau* (Bloomington: Illinois Agricultural Association, 1965), p. 13; Minutes of the Executive Committee Meeting of the DeKalb County Soil Improvement Association, January 20, 1912, DCFB Archive.

20. *DeKalb Daily Chronicle,* January 18, 30, 1912; *Sycamore True Republican,* January 20, 1912; Minutes of the Executive Committee Meeting of the DeKalb County Soil Improvement Association, January 20, 1912, DCFB Archive.

21. *DeKalb Daily Chronicle,* March 23, 1912; Minutes of the Executive Committee of the DeKalb County Soil Improvement Association, March 23, 1912, DCFB Archive.

22. *DeKalb Daily Chronicle,* March 23, 1912; Thomas H. Roberts, "Early History of the DeKalb County Soil Improvement Association, Later Renamed DeKalb County Farm Bureau" (Manuscript, 1959), DCFB Archive.

23. *DeKalb Daily Chronicle,* February 8, 1912.

24. Ibid., February 8, 28, June 4, 1912; *DeKalb County Farmer* 2:4 (Fall/November 1974): 13. The quote about corn sermons is attributed to Prof. P. G. Holden, the great corn evangelist from Iowa.

25. William G. Eckhardt, "Teaching Farmers by the DeKalb County Plan," *Prairie Farmer*, January 1, 1913; *DeKalb Daily Chronicle*, February 13, May 29, 1912, June 30, 1922. In 1912, DeKalb County had about 2,500 farms. With 700 farmer members, the Soil Improvement Association had enrolled about 28 percent of county farmers in 1912.

26. *DeKalb Daily Chronicle*, February 28, April 15, 1912.

27. Ibid., March 23, 1912, and January 17, 1913. The survey provided the data for J. G. Mosier, H. W. Stewart, E. E. DeTurk, and H. J. Snider, *DeKalb County Soils: Soil Report No. 23* (Urbana: University of Illinois Agricultural Experiment Station, 1922).

28. *DeKalb Daily Chronicle*, May 13 and 17, 1912; "Constitution and By-laws of the DeKalb County Soil Improvement Association," (1912), DCFB Archive.

29. *DeKalb Daily Chronicle*, June 1 and 4, 1912.

30. Eckhardt was an early advocate of soybeans as part of a comprehensive crop-rotation program, and had just published an article on the subject in the March, 1912, edition of *Prairie Farmer*. *DeKalb Daily Chronicle*, March 15, 1912. Soybeans were first planted in DeKalb County in 1913 and were harvested as "bean hay" for livestock.

31. *DeKalb Daily Chronicle*, June 6, 17, 24, 29, and July 8, 12, 18, 1912. *Prairie Farmer* 84:15 (August 1, 1912), p. 1. See generally, C. A. Atwood, "DeKalb County Agriculture: Ten Years Ago—Before and Since (A Sketch of the early history leading up to the establishment of Farm Bureau work in DeKalb County)," (1922), manuscript, DCFB Archive.

32. *DeKalb Daily Chronicle*, September 17, 1912.

33. Ibid., September 27, 1912; William G. Eckhardt, "How Some Concerns 'Educate' Farmers," *Prairie Farmer*, January 15, 1913. The *Chronicle* reported that Eckhardt's speech to the Banker's Association "electrified the organization."

34. *DeKalb Daily Chronicle*, July 18, August 8, and September 20, 1912.

35. Minutes of the board of directors of the DeKalb County Soil Improvement Association, July 12, 1912, DCFB Archive.

36. Minutes of the board of directors of the DeKalb County Soil Improvement Association, July 12, July 23, August 9, September 12, and October 11, 1912, DCFB Archive. The directors noted that the seed should not exceed $12 per bushel and limited the total purchase to $20,000.

37. *DeKalb Daily Chronicle*, July 24, 1912.

38. Ibid., September 14, 1912.

39. Minutes of the board of directors of the DeKalb County Soil Improvement Association, November 1912, DCFB Archive.

40. *DeKalb Daily Chronicle*, July 12, September 14, October 30, November 22, and December 14, 1912, and February 15, 1913; Thomas H. Roberts, Jr., *The Story of the DeKalb "AG"* (Carpentersville, Ill.: Carlith Printing, 1999), pp. 5–11.

41. *Breeder's Gazette*, January 22, 1913.

42. *Prairie Farmer*, August 1, 1912, and January 1, 15, 1913.

43. *Breeder's Gazette*, January 22, 1913.

44. William G. Eckhardt, "Hiring the County Expert," *Prairie Farmer*, January 1, 1913; "The Cooperative Extension Service" (Urbana: University of Illinois, circa 1962), unpublished manuscript, DCFB Archive.

3—War and Recession

1. At first, nobody was exactly certain what title Eckhardt should have. In the early days, he was variously called soil expert, soil fertility expert, agricultural advisor, agricultural expert, farm expert, soil advisor, crop doctor, demonstrator, professor (a title he held at the University of Illinois), county agent, agriculturalist, and agricultur-

ist. Most commonly, he was referred to as farm advisor. On the Soil Improvement Association letterhead, he is called "Agriculturist," the title he used when he wrote news articles, signed letters, and introduced meetings.

2. The eight counties are Kane, McHenry, DuPage, Cook, Lake, Boone, Winnebago, and Will.

3. Land prices, of course, fluctuated considerably. In 1912, farmland in Milan Township was selling for $185 to $190 per acre. Near Shabbona, a farm was advertised for $150. In June 1912, the farm realtor E. L. Larson advertised eighteen DeKalb County farms of between 80 and 320 acres for $125 to $175 per acre. The following autumn, Chris Vagel sold the 278-acre Chase Glidden farm for $200 per acre. Vagel made a $30 per acre profit on the deal after owning the Glidden farm for only one year. One 744-acre farm sold for $220 per acre. *DeKalb Daily Chronicle,* January 21, June 1, September 16, and October 25, 1912, and July 23, 1913. The average price for Illinois farmland in 1910 was $95 per acre. U.S. Department of Commerce, *Statistical Abstract of the United States, 1920* (Washington, D.C.: Government Printing Office, 1921).

4. In comparison, two years later, on January 5, 1914, Henry Ford announced that the minimum wage in his factories would be $5 per eight-hour day (for qualifying workers). It was considered at the time to be an exorbitant amount and made Ford employees among the best-paid laborers in the nation.

5. *DeKalb Daily Chronicle,* February 27, 1912.

6. *DeKalb Daily Chronicle,* September 14, 1912, and March 19, 1913.

7. Ibid., December 14, 1912.

8. "Appropriations for Soil and Crop Improvement (June 27, 1913)," *Laws of the State of Illinois, 1913* (Springfield: Illinois State Journal Co., State Printers, 1913), p. 202.

9. *DeKalb County Farmer* 1:11 (January 1, 1915), 1:12 (February 1, 1915), 2:4 (June 1, 1915), and 2:7 (September 1, 1915).

10. *DeKalb Daily Chronicle,* December 14, 1912, and March 29, 1913.

11. Ibid., March 19, August 11, and September 20, 1913; Thomas H. Roberts, "Early History of DeKalb County Soil Improvement Association, Later Renamed DeKalb County Farm Bureau" (Manuscript, 1959), p. 5, DCFB Archive; *DeKalb County Farmer* 1:4 (June 1, 1913). In 1919, 41 DeKalb farmers reported using fertilizer; ten years later, 281 farmers reported using it. During the time, there were roughly 2,000 farms in the county. U.S. Department of Commerce, Bureau of Census, *Fourteenth Census of the United States, 1920: Agriculture,* "Report for the States, The Northern States" (Washington, D.C.: Government Printing Office, 1920), p. 406; U.S. Department of Commerce, Bureau of Census, *Fifteenth Census of the United States, 1920: Agriculture,* "Report for the States, The Northern States" (Washington, D.C.: Government Printing Office, 1930), p. 649.

12. *DeKalb Daily Chronicle,* April 12, 19, July 11, 1913.

13. Ibid., September 20, November 26, 1913; William G. Eckhardt, *Illinois Farm Advisor's Annual Reports, DeKalb County, January 1, 1915 to November 30, 1919: Annual Report of Work of County Agents (1916),* p. 3, DCFB Archive.

14. *DeKalb County Farmer* 1:10 (December 1914); John J. Lacey, *Farm Bureau in Illinois: History of Illinois Farm Bureau* (Bloomington: Illinois Agricultural Association, 1965), p. 33.

15. Ibid., pp. 23–32. In 1913, Champaign, DeKalb, DuPage, Kane, Livingston, McHenry, Peoria, Tazewell, and Will counties had programs in place; by the end of 1914, Bureau, Grundy, Hancock, Iroquios, LaSalle, and Winnebago counties had them also.

16. See, generally, Robert H. Wiebe, *The Search for Order: 1877–1920* (New York: Hill and Wang, 1967).

17. The nucleus of the leadership for the Illinois Association of County Agriculturists came from four contiguous counties in northern Illinois. In addition to Readhimer and Eckhardt, the secretary-treasurer was E. B. Heaton, from DuPage County. The fourth advisor at the meeting was John Collier, of Kankakee County, who began working as farm advisor the same day as Eckhardt, June 1, 1912.

18. Edwin Bay, *The History of the National Association of County Agricultural Agents, 1915–1960* (Springfield, Ill.: Frye Printing, 1961), pp. 13–14.

19. Ibid.

20. Lacey, pp. 41–52.

21. Eckhardt, *Illinois Farm Advisor's Annual Reports, DeKalb County, January 1, 1915 to November 30, 1919: Semi-annual Report of County Agents, 1915* (June 30, 1915), p. 3, DCFB Archive.

22. Eckhardt, *Illinois Farm Advisor's Annual Reports, DeKalb County, January 1, 1915 to November 30, 1919: Annual Report of the County Agent, January 1–December 1, 1916,* p. 7, DCFB Archive.

23. Eckhardt, *Illinois Farm Advisor's Annual Reports, DeKalb County, January 1, 1915 to November 30, 1919: Report of Work of the County Agents, Calendar Year 1915,* p. 8, DCFB Archive.

24. *DeKalb County Farmer* 3:9 (November, 1916).

25. Eckhardt, *Annual Report of the County Agent, January 1–December 1, 1916,* p. 3, DCFB Archive.

26. Eckhardt, *Semi-annual Report of County Agents, 1915* (June 30, 1915), p. 11, DCFB Archive.

27. Ibid., p. 2, unnumbered supplemental page.

28. Eckhardt, *Illinois Farm Advisor's Annual Reports, DeKalb County, January 1, 1915 to November 30, 1919: Report of Work of the County Agents, Calendar Year 1915,* p. 3, DCFB Archive.

29. Ibid., pp. 10–11, DCFB Archive.

30. Minutes of the executive committee of the DeKalb County Soil Improvement Association, June 2, 1917, and minutes of the board of directors of the DeKalb County Soil Improvement Association, June 9, 1917, DCFB Archive.

31. *DeKalb County Farmer* 2:3 (May 1915), 4:4 (June 1, 1917), 4:11 (January 1, 1918); minutes of the executive committee of the DeKalb County Soil Improvement Association, June 2, 1917, and minutes of the board of directors of the DeKalb County Soil Improvement Association, June 9, 1917, DCFB Archive; *The Story of DeKalb* (DeKalb: DeKalb Agricultural Association, 1955), pp. 3–5; *DeKalb County Farmer* 3:2 (Spring/June 1975): 16. (Note that the DeKalb County Farm Bureau decided in 1972 to publish the *DeKalb County Farmer* as a quarterly and renumbered the volumes.) The Soil Improvement Association tried as early as 1914 to lease or purchase the school building "located in the east part of town" from the DeKalb School Board. Minutes of the executive committee of the DeKalb County Soil Improvement Association, October 5, 1914, DCFB Archive.

32. John T. Schlebecker, *Whereby We Thrive: A History of American Farming, 1607–1972* (Ames: Iowa State University Press, 1975), pp. 208–10; Willard W. Cochrane, *The Development of American Agriculture: A Historical Analysis* (Minneapolis: University of Minnesota Press, 1993), p. 100.

33. *DeKalb County Farmer* 4:1 (March 1, 1917).

34. David B. Danbom, *Born in the Country: A History of Rural America* (Baltimore: Johns Hopkins University Press, 1995), pp. 176–77.

35. David B. Danbom, *The Resisted Revolution: Urban America and the Industrialization of Agriculture, 1900–1930* (Ames: Iowa State University Press, 1979), p. 99.

36. Danbom, *Resisted Revolution,* pp. 99–109, and *Born in the Country,* pp. 176–80; Murray R. Benedict, *Farm Policies of the United States, 1790–1950* (New York: Octagon Books, 1966), pp. 156–68.

37. Roberts, p. 28. In the spring just after the United States declared war, corn shot to over $2.20 per bushel before settling down to between $1.70 and $1.80 for the remainder of the war.

38. *Malta Record,* April 20, 1917, *DeKalb County Farmer* 4:7 (September 1, 1917).

39. Roberts, pp. 13–14; Eckhardt, *Illinois Farm Advisor's Annual Reports, DeKalb County, January 1, 1915 to November 30, 1919: Annual Report of the County Agent, December 1, 1916 to December 1, 1917; Annual Report of the County Agent, December 1, 1917 to December 1, 1918,* DCFB Archive; *DeKalb County Farmer* 4:4 (June 1917), 4:7 (September 1917).

40. *DeKalb County Farmer* 4:7 (September 1, 1917). The early freeze destroyed seed corn that contained more than 20 percent water.

41. "Seed for the Victory Corn Crop a Work Done—A Bit of History Made," *Orange Judd Farmer,* July 13, 1918; Roberts, pp. 15–16; Eckhardt, *Annual Report of the County Agent, December 1, 1916 to December 1, 1917; Annual Report of the County Agent, December 1, 1917 to December 1, 1918,* DCFB Archive, *DeKalb County Farmer* 4:7 (September 1917), 4:10 (December 1917), 4:11 (January 1918).

42. Roberts, p. 28. Farm property tax receipts reached $619,000 in 1924.

43. Ibid. Thomas H. Roberts and Ray Nelson, *A Request for More Equitable Distribution of Taxes in DeKalb County,* September 8, 1925, DCFB Archive.

44. *DeKalb Daily Chronicle,* December 11, 1920. For a list of farm product prices, see, generally, Harold Barger and Hans L. Landsberg, *American Agriculture, 1899–1939* (New York: National Bureau of Economic Research, 1942), pp. 332–52.

45. See, for example: H. Thomas Johnson, *Agricultural Depression in the 1920s: Economic Fact or Statistical Artifact?* (New York: Garland, 1985).

46. Danbom, *Born in the Country,* pp. 185–205; Benedict, pp. 173–238.

47. Bureau membership was 2,024 out of a total of 2,481 farms in the county. "Farm Bureau Historical Pageant (Eckhardt's and Robert's Administration)," text of speech for the 1922 historical pageant (no author, n.d.), DCFB Archive.

48. The American Farm Bureau Federation was organized in 1919. Membership in county farm bureaus grew tremendously during the war, and they evolved into something more than simply providers of agricultural education. By the end of the war, American conservatives looked to the growing network of farm bureaus as counterweights to bolshevism and radical domestic farm movements, especially the left-leaning Nonpartisan League. The bureaus, after all, were closely allied with the agricultural business community and were steeped in traditional American ideals including self-reliance, market capitalism, Puritan morality, and democracy. At the organizing meeting in Chicago in November 1919, 500 representatives from thirty-six states (half the delegates were from Illinois) voted to create a "sane" farm organization. See Lowell K. Dyson, *Farmers' Organizations* (New York: Greenwood Press, 1986), pp. 14–23; Edward L. Schapsmeier, *Henry A. Wallace of Iowa* (Ames: Iowa State University Press, 1968).

49. Harriet Wilson Davy, *From Oxen to Jets: A History of DeKalb County, 1835–1963* (DeKalb, Ill.: DeKalb County Board of Supervisors, 1963), pp. 45–46; *DeKalb Daily Chronicle,* April 21 and 30, May 3, and June 2, 1920; Lacey, pp. 72–76. The Committee of Seventeen's efforts resulted in the creation of the United States Grain Growers, in 1921. Although nearly 50,000 farmers joined the new organization, poor business practices and internal policy struggles within the American Farm Bureau Federation killed it. By 1922, it was a quarter million dollars in debt and went out of business in 1923. In 1923, Eckhardt sued Professor J. Clyde Filley and the Ne-

braska Farm Bureau Federation for libel. Filley had published an article in the Nebraska Farm Bureau newsletter claiming, among other things, that the United States Grain Growers was bankrupt in March 1922, that the Grain Growers made financial mistakes and spent money unwisely, that Eckhardt was unfit for the position as treasurer of the organization, and generally that Eckhardt was a leader of the "radical group of the United States Grain Growers' Association." Eckhardt lost his suit. Harry Keefe, attorney for the Farm Bureau, said, "Eckhardt was a man possessed of a great vision as was shown by his activities as advisor to the farmers in Illinois but . . . he was not practical in his views." *DeKalb Daily Chronicle,* November 6, 7, 8, and 13, 1923; Eduard C. Lindeman, "Social Implications of the Farmers Cooperative Marketing Movement," *Journal of Social Forces* 1:4 (1923). Eckhardt returned to DeKalb County and opened a private seed business. He also developed an improved corn drier, the Eckhardt Ear Corn Drier, that was "the greatest step forward during the past 50 years in saving and making the greatest possible use of the corn crop for feed and seed." *DeKalb Daily Chronicle,* April 11, 1924.

50. Thomas H. Roberts, *Annual Report of the County Farm Advisor, December 1, 1919 to November 30, 1920,* DCFB Archive.

51. Even in the peak income tax years during World War I, 95 percent of earners were exempt from tax. By the end of the 1920s, 80 percent of earners were exempt. The fact that DeKalb County had at least 300 individuals who had to file income tax returns illustrates how prosperous the county was even in difficult economic times.

52. *DeKalb Daily Chronicle,* January 2 and January 27, 1920; Roberts, *Annual Report of the County Farm Advisor, December 1, 1919 to November 30, 1920; Annual Report of the County Farm Advisor, December 1, 1920 to November 30, 1921,* DCFB Archive.

53. Sara Buerer Mendez, *Wigwams to Moon Footprints* (Waterman, Ill.: Waterman Press, 1972), pp. 69–70; U.S. Department of Commerce, Bureau of the Census, *United States Census of Agriculture, 1925,* p. viii; U.S. Department of Commerce, *Fifteenth Census,* p. 649. Altogether, farmers in DeKalb County cooperatively sold more than $895,000 of products, and bought more than $62,000, in 1924, the peak year for cooperatives throughout the county. Even as economic conditions in rural America improved late in the decade, DeKalb County farmers were still interested in their cooperatives—in 1929 they sold $785,500 worth of products cooperatively and bought $68,000. See, generally, Henry H. Bakken and Marvin A. Schaars, *The Economics of Cooperative Marketing* (New York: McGraw-Hill, 1937).

54. Roberts, *Annual Report of the County Farm Advisor, December 1, 1920 to November 30, 1921.*

55. Roberts, *Annual Report of the County Farm Advisor, December 1, 1919–November 30, 1920; Annual Report, December 1, 1920 to November 30, 1921; Annual Report, July 1, 1922 to June 30, 1923;* minutes of directors meeting of the DeKalb County Soil Improvement Association, January 14, 1922, DCFB Archive. Although the term "Federal" appears in the name, Federal Land Banks did not rely on federal funds. They were farmer-owned cooperative associations that sold bonds to raise revenue which they used to make loans.

56. William G. Eckhardt, "The Federal Land Bank," DeKalb Historical Collection, Northern Illinois Regional History Center, Northern Illinois University, DeKalb, Illinois. Despite the new financing opportunities, land tenancy in DeKalb rose during the 1920s, from about 51 percent in 1920 to nearly 56 percent by 1930, while the average size of farms in the county increased. The size of an average farm in DeKalb County in 1920 was about 158 acres, 160 acres in 1925, and 164 acres in 1930. The rise in tenancy is most likely explained by county farmers who slipped into bankruptcy and remained as tenants on land they once owned. Others simply sold their

holdings to their more successful neighbors and ceased farming altogether. See U.S. Department of Commerce, *Fifteenth Census of the United States, 1930*, p. 649.

57. Roberts, *Annual Report of the Couny Farm Advisor, December 1, 1919 to November 30, 1920; Annual Report of the County Farm Advisor, December 1, 1920 to November 30, 1921*, DCFB Archive.

58. Roberts, *Illinois Farm Advisor's Annual Report, 1922*, pp. 1–5; minutes of the board of directors meeting, DeKalb County Soil Improvement Association, January 14, 1922, DCFB Archive.

59. Letter, G. E. Metzger to Farm Bureau President and Executive Committee (n.d., c. 1922); minutes of the board of directors meeting, DeKalb County Farm Bureau, April 28, 1922, DCFB Archive.

60. *DeKalb Daily Chronicle*, January 9, 12, and 26, March 15 and 27, April 26, May 15, 19, and 27, June 15, 22, 24, 29, and 30, and July 1, 1922.

61. Ibid., June 30, 1922.

62. *DeKalb County Farmer* 5:7 (May 1922); "Farm Bureau Decennial Celebration," June 30, 1922, DCFB Archive.

63. *DeKalb Daily Chronicle*, September 6, 1922.

64. For example, in 1925 there were 48,300 head of cattle in DeKalb County, with a total value of $2,932,000, or about $60.70 per head. Four years later, in 1929, county farmers raised 42,500 head, with a total value of $3,643,000, or about $85.74 per head. William J. Strateon, ed., *Blue Book of the State of Illinois, 1929–1930* (Springfield, Ill.: Journal Publishing, 1929), pp. 516–20; "County Estimates for Twenty Years Ending January 1, 1945," *Illinois Agricultural Statistics* (Springfield: Illinois Cooperative Crop Reporting Service, 1951), pp. 236–38. While corn was the principal grain crop, DeKalb County farmers also grew a considerable amount of oats for livestock and draft animals. "State Data through 1944," *Illinois Agricultural Statistics* (Springfield: Illinois Cooperative Crop Reporting Service, 1949), p. 11; *DeKalb Daily Chronicle*, February 7, 1972.

65. Minutes of the board of directors meeting of the DeKalb County Agricultural Association, April 28, 1922; minutes of the board of directors meeting of the DeKalb County Soil Improvement Association, December 15, 1923, DCFB Archive.

66. Roberts, *Illinois Farm Advisor's Annual Report, December 1, 1923 to November 30, 1924; Annual Report, December 1, 1926 to November 30, 1927*; minutes of the executive committee meeting, DeKalb County Farm Bureau, September 28, 1927; minutes of the board of directors meeting, DeKalb County Farm Bureau, December 17, 1927, DCFB Archive.

67. Roberts, *Illinois Farm Advisor's Annual Report (1923)* (Narrative Report), pp. 5–6; See, generally, *Farm Advisor's Monthly Narratives* (1925–1928); minutes of the meeting of the board of directors, DeKalb County Soil Improvement Association, December 15, 1923; minutes of the executive committee of the DeKalb County Farm Bureau, September 28, 1927, DCFB Archive.

68. Roberts, *Annual Report of the County Extension Agent, November 30, 1923 to November 30, 1924; Annual Report, 1925; Annual Report, 1926*. See, generally, *Farm Advisor's Monthly Narratives* (1925–1928), DCFB Archive. Also see, generally, Thomas Wessel and Marilyn Wessel, *4-H, An American Idea, 1900–1980: A History of 4-H* (Chevy Chase, Md.: National 4-H Council, 1982).

69. Minutes of the board of directors meeting, DeKalb County Soil Improvement Association, June 9, 1925, and December 17, 1927, DCFB Archive.

70. Henry C. Taylor, Anne D. Taylor, and Norman J. Wall, *The Story of Agricultural Economics in the United States, 1840–1932* (Ames: Iowa State College Press, 1952), pp. 996–97; *DeKalb Daily Chronicle*, December 8, 1928.

71. *DeKalb Daily Chronicle*, April 28, 1920; Thomas H. Roberts and Raymond C. Nelson, "A Request for More Equitable Distribution of Taxes in DeKalb County,"

(DeKalb County Soil Improvement Association, September 8, 1925); Roberts, "Early History," pp. 24–26; Roberts, *Illinois Farm Advisor's Annual Report, December 1, 1923 to November 30, 1924; Illinois Farm Advisor's Annual Report, December 1, 1925 to November 30, 1926; Annual Report, December 1, 1926 to November 30, 1927; Farm Advisor's Monthly Narratives* (September 1925); minutes of the board of directors of the DeKalb County Soil Improvement Association, June 9, 1925, DCFB Archive. *DeKalb County Yearbook,* 1924–1929, Illinois Regional Archives Depository, Northern Illinois University, DeKalb, Illinois (hereinafter, IRAD). The tax-equalization battle was carried on at the state level by the Illinois Agricultural Association, the state parent organization of Illinois Farm Bureaus. The Illinois Agricultural Association campaign began in 1921 and targeted the Illinois Tax Commission. In 1919, Illinois railroads, led by the Chicago and Northwestern Railway, had presented statistics showing that Illinois farmland was underassessed. To contradict that evidence, the Illinois Agricultural Association presented its own evidence to the commission in 1922 that showed that tax rates for railroads had risen at a slower rate than for farmland and that farmland was generally overassessed in relation to its value. The tax commission was persuaded by the Illinois Agricultural Association evidence and reduced taxes on agricultural land statewide. See Taylor, Taylor, and Wall, pp. 997–98.

In DeKalb County, total tax assessments increased from $31,594,000 in 1920 to $52,321,000 in 1928. Agricultural tax assessments, however, increased only from $17,580,000 to $27,281,000 during the same time. In other words, agricultural assessments, as a percentage of total county tax assessments, fell from about 55.6 percent in 1920 to 52.1 percent in 1928. The high point was 1924, just before the tax relief program went into effect, when agricultural assessments were 56.4 percent of total assessments. *DeKalb County Yearbook,* 1921–1929; DeKalb County, *Statement of Assessments,* 1920–1928, IRAD.

72. Minutes of the Soil Improvement Association meeting, April 28, 1929, DCFB Archive.

73. Lacey, pp. 100–101, 123–24.

74. Roberts, "Early History," p. 21; Thomas H. Roberts, *Illinois Farm Advisor's Annual Report* (1927). See, generally, *Farm Advisor's Monthly Narratives* (1925–1928), DCFB Archive. Lacy, pp. 100–101, 123–24.

75. Minutes of the board of directors meeting, DeKalb County Farm Bureau, March 25 and December 17, 1927, DCFB Archive. There were roughly 158,000 tractors in the United States in 1919, and 827,000 ten years later. Harold Barger and Hans L. Landesberg, *American Agriculture, 1899–1939* (New York: National Bureau of Economic Research, 1942), p. 204.

76. Letter, Thomas Roberts to "Patron," September 10, 1929, Box 1, RC 52, Northern Illinois Regional History Center, Northern Illinois University, DeKalb, Illinois; Edna Nelson, "A Report on Kishwaukee Service Company," May, 1962, DCFB Archive. The delivery trucks were numbered, and originally each one served a designated part of the county—truck no. 1 served the DeKalb area, no. 2 for Hinckley, no. 3 for Waterman, no. 4 for Sandwich and Somonauk, no. 5 for Genoa, and no. 6 for Sycamore. Ibid.

4—Hard Times

1. David B. Danbom, *Born in the Country: A History of Rural America* (Baltimore: Johns Hopkins University Press, 1995), p. 196.

2. Reynold M. Wik, "The Radio in Rural America During the 1920s," *Agricultural History* 55 (October 1981): 339–50.

3. Danbom, *Born in the Country,* pp. 185–97, and *The Resisted Revolution: Urban America and the Industrialization of Agriculture, 1900–1930* (Ames: Iowa State University Press, 1979), pp. 120–28; Sidney Glazer, "The Rural Community in the Urban Age," *Agricultural History* 23:2 (April 1949).

4. U.S. Federal Farm Board, *First Annual Report for the Year Ending June 30, 1930* (Washington, D.C.: Government Printing Office, 1930), pp. 64–70.

5. Danbom, *Born in the Country,* pp. 197–200; U.S. Department of Agriculture, *Yearbook of Agriculture, 1935* (Washington, D.C.: Government Printing Office, 1935), pp. 673–86.

6. Lowell K. Dyson, *Farmers' Organizations* (New York: Greenwood Press, 1986), pp. 83–90. See, generally, Studs Terkel, *Hard Times* (New York: Pantheon Books, 1986). In DeKalb County, the *DeKalb County Farmer* offered advice to farmers about their horses through the late 1930s. See, for example, *DeKalb County Farmer* 21:6 (June 1933), 23:11 (November 1935), and 26:6–7 (June–July 1938).

7. Author's interview with Wayne Bark, May 15, 2003.

8. *DeKalb Daily Chronicle,* October 28 and November 19, 1928.

9. See, generally, Thomas H. Roberts, *Annual Report of the DeKalb County Farm Bureau,* 1920–1929; *Farm Advisor's Monthly Narratives,* July 1925–March 1928, DCFB Archive.

10. Orville M. Kline, *The Farm Bureau Movement* (New York: Macmillan, 1921), pp. 107–8.

11. Minutes of the executive Committee meeting of the DeKalb County Farm Bureau, June 29, 1927, DCFB Archive.

12. Letter, William Brown to E. E. Golden, June 13, 1973, DCFB Archive.

13. Roberts tendered his resignation and accepted a position as manager of the Sycamore Preserve Works, a facility that canned locally grown vegetables. The DeKalb County Agricultural Association hired Roberts shortly afterward. *DeKalb Daily Chronicle,* April 4, 1930.

14. McHenry, Henry, LaSalle, McLean, and Bureau counties all had higher values. The total value of DeKalb county livestock was estimated at $4,713,300, and the total crop value was estimated at $6,335,140. *DeKalb Daily Chronicle,* March 19, June 24, and September 16, 1931.

15. "Farm Bureau Historical Pageant (Eckhardt's and Robert's Administration)," text of speech for the 1922 historical pageant (no author, n.d.), DCFB Archive. See, generally, A. Richard Crabb, *Hybrid Corn Growers: Prophets of Plenty* (New Brunswick, N.J.: Rutgers University Press, 1948).

16. In 1935 there was a waiting list for the Association's hybrid corn seed, but the list reflected more the extremely limited amount of the seed than widespread demand for the product. *DeKalb County Farmer* 23:3 and 23:11 (March and November 1935).

17. Author's interview with Ray Larson, May 8, 2003.

18. DeKalb County Agricultural Association, *The Story of DeKalb* (DeKalb, Ill.: DeKalb County Agricultural Association, 1955). See, generally, Thomas H. Roberts, Jr., *The Story of the DeKalb "AG"* (Carpentersville, Ill.: Carlith Printing, 1999); *DeKalb County Farmer,* 21:2 (February 1933), 24:1 (January 1936); *DeKalb County Farmer* 1:4 (Fall/November 1973): 10–12.

One area of social research into rural communities is the issue of diffusion of knowledge and innovation. One study finds that it took about fourteen years from the time hybrid seed was first available for it to become widely accepted. More detailed analysis shows that communities had a few innovators who tried it first, while others waited to see what happened. A second group tried the new seed only after they saw how it worked for the innovators. Perhaps the most interesting aspect of this study is that the majority of farmers did not look to the first innovators as role models, but rather to those who adopted hybrid technology somewhat later. In other words, in rural communities during the first half of the century, there was a hierarchy of influence that worked to delay, not accelerate, the adoption of new technologies. Don A. Dillman, "The Social Environment of Agriculture and Rural Communities," in R. J. Hildreth, Kathryn L. Lipton, Kenneth C. Clayton, and Carl C. O'Connor, eds., *Agriculture and Rural Areas Approaching the Twenty-First Century* (Ames: Iowa State University Press, 1988), p. 65.

19. Letter, Brown to Golden.

20. See, generally, *DeKalb County Farmer,* 1932–1935.

21. Russel N. Rasmusen, "Farm Bureau Activities from August 1929 to January, 1937" (undated manuscript), DCFB Archive. Author's interview with Joseph Faivre, March 18, 2003; Author's interview with George Tindall, May 11, 2003. In 1935, future DeKalb County Farm Bureau president and national farm policy leader Howard Mullins won the state first prize in Chicago with his 4-H steer. *DeKalb County Farmer* 23:4 (April 1935).

22. *DeKalb County Farmer* 19:4 (April 1931), 23:6 (June 1935), 24:7 (July 1936), 25:8 (August 1937), 26:4 and 9 (April and September 1938), 27:10–11 (October–November 1939), 28:9–10 (September–October 1940); author's interview with George Tindall. Probably the first purely social function sponsored by the Farm Bureau was the Farmers' Dinner Club. Neighboring Winnebago County had such a club, and Roberts told the DeKalb Bureau's board of directors that he could secure Colonel Clarence Ousley, of Dallas, Texas, as the speaker for the first farmers' dinner, held December 31, 1927. Minutes of the board of directors meeting of the DeKalb County Farm Bureau, December 17, 1927, DCFB Archive.

23. Minutes of the directors meeting, DeKalb County Agricultural Association, April 28, 1922, DCFB Archive. *DeKalb County Farmer* 24:9 (September 1936).

24. *DeKalb County Farmer* 24:10 (October 1936), 25:1 and 4–5 (January and April–May, 1937). Biester was succeeded by Mrs. Venus Johnson Evans, in early 1939. Evans turned over the home advisor duties to Helen Johnson, of Sandwich, in May 1940. *DeKalb County Farmer* 28:5 (May 1940).

25. DeKalb County Home Bureau, "Scrapbook, 1939–1940, Other Early Years Included." DeKalb County Extension Office, DeKalb, Illinois. Author's interview with Jeff and Mary Lu Strack, July 22, 2003. *DeKalb County Farmer* 25:7–10 (July–October 1937), 26:1 (January 1938).

26. Rasmusen, "Farm Bureau Activities from August 1929 to January, 1937."

27. Minutes of the executive committee meeting, DeKalb County Farm Bureau, September 19, 1936; minutes of the board of directors meeting, DeKalb County Farm Bureau, November 7 and December 22, 1936, DCFB Archive. *DeKalb County Farmer* 24:5 (May 1936), 24:10 (October 1936).

28. Edna Nelson, "A Report on Kishwaukee Service Company," May 1962, DCFB Archive. Nelson states that the name change was finalized July 1, 1938. A. E. Kiefer and J. O. Matthewson were named the Bureau's representatives on the new board of directors for the oil service company. Minutes of the board of directors meeting, DeKalb County Farm Bureau, November 7, 1936, DCFB Archive. At the end of 1944, the farm bureaus of Boone and McHenry counties left Kishwaukee Service Company to form their own oil delivery company, just as the Ogle County Farm Bureau had done a few years earlier. That same year, the Kishwaukee Service Company retired all the remaining capital stock held by the DeKalb County Agricultural Association. It continued to have a close relationship with the Farm Bureau and offer Bureau members special rates as one of the Bureau's affiliated companies. J. O. Matthewson, "Minutes of Joint Meeting of the Directors of Kishwaukee Service Company," October 27 and November 2, 1944, DCFB Archive.

29. "Supplement to the report of the President to the 1938 annual meeting, etc." (no author), 1938, DCFB Archive.

30. Letter, A. E. Keifer and J. O. Matthewson, Rural Improvement Society, to DeKalb County Farm Bureau, February 15, 1939. Minutes of the board of directors meeting, DeKalb County Farm Bureau, March 3, 1939, DCFB Archive.

31. Minutes of the executive committee meeting, DeKalb County Farm Bureau, November 30, December 22, 1936; "Supplement to the report of the President to the 1938 annual meeting, etc.," 1938, DCFB Archive. The Rural Improvement Society consisted of ten members. Seven memberships are selected by the Farm Bureau executive committee or the Farm Bureau board of directors. Historically, however, the Farm Bureau secretary holds seven proxy votes, and the other three votes are controlled by the Farm Bureau president, vice president, and treasurer—four people who are Farm Bureau members control all ten votes of the Society.

32. The Farm Bureau had assisted farmers with farm accounting as early as 1915, but especially since 1925. *DeKalb County Farmer* 23:11 (November 1935), 24:1 (January 1936).

33. Minutes of the board of directors meeting of the DeKalb County Farm Bureau, March 3, 1939, DCFB Archive. *DeKalb County Farmer* 25:4–9 (April–September 1937), 26:4 and 7 (April and July 1938), 27:2–3 (February–March 1939). Locker Service stock was originally offered at $25 per share, with a 6 percent annual yield. The Locker Service facility in DeKalb opened in January 1938. By March, the Sycamore facility was filled to capacity. In May, plants opened in Waterman and Somonauk. By March 1939, the Bureau had $1,500 invested in the five company plants in the county. *DeKalb County Farmer* 26:1–5 (January–May 1938).

34. *DeKalb County Farmer* 21:4 (April 1933) and 21:5 (May 1933).

35. P. E. Johnson, J. B. Cunningham, and E. M. Hughes, *Farm Business Report, 1940: Farming-Type Area Two, Northwestern Mixed Livestock Area* (Urbana: University of Illinois College of Agriculture, Extension Service, 1940).

36. John T. Schlebecker, *Whereby We Thrive: A History of American Farming, 1607–1972* (Ames: Iowa State University Press, 1975), p. 213; P. E. Johnson, J. B. Cunningham, and W. D. Buddemeier, *Farm Business Report, 1941: Farming-Type Area Two, Northwestern Mixed Livestock Area* (Urbana: University of Illinois College of Agriculture, Extension Service, 1941).

37. Johnson et al., *Farm Business Report, 1940;* J. B. Cunningham, E. N. Searls, and O. B. Brown, *Farm Business Report, 1944: Farming-Type Area Two, Northwestern Mixed Livestock Area* (Urbana: University of Illinois College of Agriculture, Extension Service, 1944). Total wheat production nationally, for example, increased nearly 50 percent between 1939 and 1945. In northwestern Illinois, farmers continued to produce corn at the same maximum rates they had during the late 1930s. Average corn yields increased only about a bushel per acre, to sixty-five bushels, between 1940 and 1944.

38. *DeKalb County Farmer* 30:8–11 (August–November 1942).

39. The Illinois Agricultural Association reported that in May 1942 farm prices were 18 percent less than prices in May 1917. For example, farmers earned $0.81 per bushel for corn in May 1942 compared with $1.55 in May 1917; a price of $1.00 per bushel for wheat in May 1942 compared with $2.47 per bushel in May 1917. Eggs sold for $0.26 per dozen in 1942, but $0.30 in 1917. Lambs sold for $11.62 per hundred pounds in 1942, but $12.51 in 1917. *DeKalb County Farmer* 30:6 (June 1942); J. B. Cunningham, E. L. Sauer, and E. N. Searls, *Farm Business Report, 1943: Farming-Type Area Two, Northwestern Mixed Livestock Area* (Urbana: University of Illinois College of Agriculture, Extension Service, 1943); Cunningham et al., *Farm Business Report, 1944.*

40. *DeKalb County Farmer* 29:11 (November 1941), 30:2 (February 1942); Cunningham et al., *Farm Business Report, 1944.*

41. Walter W. Wilcox, *The Farmer in the Second World War* (Ames: Iowa State College Press, 1947), pp. 308–9; Murray R. Benedict, *Farm Policies of the United States, 1790–1950* (New York: Octagon Books, 1966), p. 453. The DeKalb County Farm Bureau actively worked to persuade its members to use their newfound wealth to pay debts and buy savings bonds. See, for example, *DeKalb County Farmer* 30:2 and 3 (February and March 1942).

42. "Annual Report of the DeKalb County Farm Bureau and Farm Bureau Associated Companies and Other Agricultural Agencies," 1941; "Who's Who in Farm Bureau: Historical Data" (n.d.); Al Golden interview of Donald Mosher, August 28, 1979, DCFB Archive.

43. *DeKalb County Farmer* 30:9 (September 1942).

44. *DeKalb County Farmer* 30:8 (August 1942); "Program and Reports, 34th Annual Meeting of DeKalb County Farm Bureau and 8th Annual Meeting of Kishwaukee Service Company," February 5, 1946, DCFB Archive.

45. The Kishwaukee Service Company rubber drive in the spring of 1942, for example, netted 26.2 tons of rubber from its customers. *DeKalb County Farmer* 30:6 (July 1942), 31:10 (October 1943), 32:6 (June 1944). Oil rationing was only one problem—a shortage of steel drums led the Kishwaukee Service Company to caution farmers about how they cared for them. *DeKalb County Farmer* 30:7 (July 1942), 31:2 (February 1943); minutes of the board of directors, DeKalb County Farm Bureau, May 22, 1945, DCFB Archive.

46. See, generally, *DeKalb County Farmer* 30:1–12 (January–December 1942), 31:1–12 (January–December 1943), 32:1–12 (January–December 1944). Author interview with George Tindall. "Program and Reports, 34th Annual Meeting of DeKalb County Farm Bureau and 8th Annual Meeting of Kishwaukee Service Company," February 5, 1946, DCFB Archive.

47. *DeKalb County Farmer* 30:2 (January 1942), 31:4 (April 1943). Farm Bureau leaders persuaded the army to release Johnson from duty so that he could return to his position as farm advisor. Johnson refused the release. "I didn't feel that I as an individual had the right to ask for anything that had not been previously specified by Extension Service," wrote Johnson. "I did not want my mother to have to hear occasional comment that her son was released because he had influence. I did not want someone else, perhaps even someone with dependents, to have to take my place in the armed ranks. . . . I fully realize what I have given up in making my decision. I could not have asked for better employers or for finer working conditions. I thank each of you for your patience and co-operation." Letter, Roy Johnson to E. E. Houghtby (n.d.), read at Board of Directors' Meeting, DeKalb County Farm Bureau, May 26, 1942, DCFB Archive.

48. *DeKalb County Farmer* 30:12 (December 1942), 31:1–4 (January–April 1943), 32:6 (June 1944). Although hemp farming did not last long, hemp continued to grow wild along ditch banks and fencerows for years after the war ended.

49. *DeKalb County Farmer* 31:11 (November 1943).

50. In late 1944, the DeKalb County Agricultural Association held $65,000 of first preferred Kishwaukee Service Company stock. Kishwaukee Service had $47,000 of cash reserve. In October, the Kishwaukee Service Company, which sold petroleum and other goods in DeKalb, McHenry, and Boone counties, agreed to dissolve into three separate companies representing each county. The executive committee of the Farm Bureau agreed to retain the Kishwaukee Service structure, and the other counties created new corporate entities. All three companies would be members of the Illinois Farm Supply Company, the statewide cooperative that provided petroleum and other

products to its member companies. Assets and liabilities were divided in proportion to the amounts applying in and to each county. Kishwaukee Service used its cash reserves to retire the Association's preferred stock. In early 1945, the Farm Bureau purchased $5,000 of class A stock, 200 shares, in the reorganized Kishwaukee Service Company and recommended to the Kishwaukee board of directors that they offer the remaining 2,400 shares only to Farm Bureau members in the county, at a par value of $25.00, with no more than four shares sold to any one person. Interest was set at 5 percent. Edna Nelson, "A Report on Kishwaukee Service Company," May, 1962; minutes of joint meeting of the directors of DeKalb County Farm Bureau and the newly elected directors of McHenry County Service Company and Boone County Service Company, etc., October 27, 1944, November 2, 1944; "Suggested Basis of Contract as Determined by Committee Nov. 2, 1944," (n.d.); Minutes of the Board of Directors Meeting (of the Farm Bureau)," January 23, 1945; minutes of "Special Meeting," February 1, 1945, DCFB Archive.

5—Postwar Years

1. Letter, L. D. Sears to Mr. Louis Lloyd, May 27, 1939, DCFB Archive. Parke, a humble man by reputation, preferred to characterize the gathering as the anniversary of the Smith-Lever Act rather than a banquet held in his honor. Letter, H. H. Parke to Mr. Louis Lloyd, June 12, 1939, DCFB Archive.

2. E. E. Golden, "Data Relating to the Farm Problem" (DeKalb, Ill.: Cooperative Extension Service, DeKalb County, 1960), DCFB Archive; J. B. Cunningham, E. N. Searls, and O. B. Brown, *Farm Business Report, 1944: Farming-Type Area Two, Northwestern Mixed Livestock Area* (Urbana: University of Illinois College of Agriculture, Extension Service, 1944).

3. Russel Rasmusen, "Farm Bureau Activities from August 1929 to January 1937," (n.d.), DCFB Archive. See, generally, Deborah K. Fitzgerald, *The Business of Breeding: Hybrid Corn in Illinois, 1890–1940* (Ithaca, N.Y.: Cornell University Press, 1990).

4. David Pimentel, L. E. Hurd, A. C. Bellotti, M. J. Foster, I. N. Oka, O. D. Sholes, and R. S. Whitman, "Food Production and the Energy Crisis," *Science* 182 (November 2, 1973): 443–49; G. H. Heichel, *Comparative Efficiency of Energy Use in Crop Production,* Bulletin 739, Connecticut Agricultural Experiment Station, New Haven, November 1973.

5. Although it is true that industrial wages increased faster than net farm income, the contrast may not be as great as statistics seem to show because nonfarm expenses also increased faster than farm expenses. For example, the cost of living for farmers increased about 20 percent between 1947 and 1959, but urban cost of living jumped 24 percent. Urban retail food costs rose 18 percent. Likewise, farm machinery costs increased 55 percent during the same time, whereas manufacturing component costs increased 52 percent. Moreover, although farmers continued to rely on other suppliers for much of the raw materials of farming—seed, machinery, finished goods, capital, and so forth—the farm also continued to provide staples for many farm families that were not available to urban folks. Golden, "Data Relating to the Farm Problem."

6. Ibid.

7. Ibid.

8. Henry Parke, "What Was, What Is and What Will Be," Address to the Sycamore Rotary Club, 1954, DCFB Archive.

9. Willard W. Cochrane and Mary E. Ryan, *American Farm Policy, 1948–1973* (Minneapolis: University of Minnesota Press, 1976), pp. 1–20. Parke, "What Was, What Is and What Will Be."

10. "DeKalb County, Illinois Farm Bureau and Subsidiaries" (no author, 1946), DCFB Archive.

11. By the end of the 1950s, the insurance program even offered a "dread disease" policy that provided families with security from medical expenses incurred by diphtheria, scarlet fever, rabies, smallpox, leukemia, tetanus, encephalitis, tularemia, spinal meningitis, polio, cancer, and heart disease. See, for example, DeKalb County Farm Bureau, "Annual Report, 1957," DCFB Archive.

12. *DeKalb County Farmer* 37:2 (February, 1950); "DeKalb County, Illinois Farm Bureau and Subsidiaries." At the end of the war, the DeKalb County Farm Bureau was affiliated with and supported DeKalb County Producers Supply, DeKalb County Home Bureau, Farm Bureau Insurance Company, Kishwaukee Service Company, DeKalb County Locker Service, Production Credit Association, DeKalb County A.A.A., and the U.S. Soil Conservation Service. It also continued its close, informal association with the DeKalb County Agricultural Association. See, generally, "Annual Report, DeKalb County Farm Bureau and Associated Companies," 1944–1950, DCFB Archive. The DeKalb County Grain Company was authorized to issue 3,000 shares of class A preferred stock at $100 par value with 6 percent dividend, 3,000 shares of common stock with no par value, and 6,000 shares of class B stock with no par value issued to the Illinois Grain Terminals Company. Ownership of common stock and patronage dividends was available only to Farm Bureau members. A majority of the board of directors of the grain company was required to be members of the DeKalb County Farm Bureau board. The Farm Bureau also loaned the grain company $15,000 to cover immediate start-up expenses. A second stock offering in 1950 raised an additional $60,000. The company bought the elevator in Somonauk, which had a 60,000-bushel storage capacity and a grinding facility. The merchandising was done primarily through the "Ironside Elevator" owned by the Illinois Grain Terminals Company in Chicago. In 1960, the DeKalb County Grain Company merged with the Kishwaukee Service Company. Minutes of the board of directors meeting of the DeKalb County Farm Bureau, May 19, 1949; minutes of the executive committee of the DeKalb County Farm Bureau, June 21, 1949; *DeKalb County Farmer* 37:8 (August 1949), 38:10 (October 1950); articles of merger of DeKalb County Grain Company and Kishwaukee Service Company, June 3, 1960, DCFB Archive.

13. See, generally, *DeKalb County Farmer;* DeKalb County Farm Bureau annual reports, 1942–1950, DCFB Archive. One of the important services that the Farm Bureau provided to 4-H youth was that the Bureau's livestock committee secured calves for livestock projects.

14. DeKalb County Farm Bureau, "Annual Report: DeKalb County Farm Bureau and Associated Companies," 1947, 1949, DCFB Archive. Thomas H. Roberts, Jr., *The Story of the DeKalb "AG"* (Carpentersville, Ill.: Carlith Printing, 1999), p. 74.

15. During the war, the Farm Bureau invested some of its "building fund" in U.S. War Bonds. Minutes of the board of directors meeting, DeKalb County Farm Bureau, May 22, 1945, DCFB Archive.

16. DeKalb County Farm Bureau, "Open House (Building Dedication Program), August 29, 1951," DCFB Archive.

17. The original contractors' bids totaled $197,489.29. Minutes of the executive committee of the DeKalb County Farm Bureau, April 25, 1950; minutes of the board of directors meeting of the DeKalb County Farm Bureau, May 23, 1950, DCFB Archive.

18. Minutes of the executive committee meeting of the DeKalb County Farm Bureau, March 27, 1945, DCFB Archive. Assistant advisor Roy E. Will resigned in August 1951. Advisor Ward Cross resigned in February 1952. Assistant Cliff Heaton was promoted to acting farm advisor, and Ralph Stock was hired in May 1952, as Heaton's new assistant. Both men left in the spring of 1954. The Bureau hired E. E. Golden as its ninth farm advisor in May 1954.

19. Letter, William Brown to E. E. Golden, June 13, 1973, DCFB Archive.

20. Author's interview with E. E. Golden, May 22, 2002.

21. Author's interview with Howard Mullins, July 9, 2003.

22. DeKalb County Farm Bureau, annual reports, 1957, 1958, and 1959, DCFB Archive. In June 1956, the DeKalb County Farm Bureau float won first place in the DeKalb Centennial Parade. It featured Howard Mullins and a photograph of the Farm Bureau's new building. Sitting at the front of the float in a throne embossed with "Farm Bureau Originator," sat Henry Parke. Less than a year later, on May 25, 1957, Parke died, breaking the Bureau's last link with the founding fathers of the Soil Improvement Association. *DeKalb County Farmer* 33:8 (August 1956), 34:7 (July 1957).

23. See, generally, *DeKalb County Farm Bureau Annual Report, 1956–1960*, DCFB Archive. *DeKalb County Farmer*, 1950–1960. The success of cooperative marketing, however, convinced the Bureau that consolidating their operations fully with the state Illinois Livestock Marketing Association would offer even greater benefits to its members and so merged the Bureau's operations with the state's completely in 1960.

24. *DeKalb County Farmer* 35:9 (September 1958).

25. See, generally, *DeKalb County Farmer*, 1957–1958.

26. "Report of Farm Policy Council Meeting at Des Moines, Iowa, December 8–9, 1960," DCFB Archive.

27. Mullins's informal advisors included Tom Roberts, Sr., Russel Rasmusen, both former farm advisors, Montavon, and farm advisor E. E. Golden.

28. J. Carroll Bottum, "Illinois Farmers' Farm Program," May 9, 1960, DCFB Archive.

29. Minutes, "Meeting of Farm Policy Council Held in Farm Bureau Building, DeKalb Illinois, Thursday, May 26, 1960"; Farm Policy Council, press release, "Formation of Farm Policy Council Announced" (n.d.); statement of "U.S. Farm Policy Foundation" (n.d.); U.S. Farm Policy Council, "Statement of Functions and Objectives," (n.d.); letter, Montavon and Mullins to Bernard Collins, May 13, 1960; letter, Montavon and Mullins to John Chamberlain, May 18, 1960; letter, Montavon and Mullins to "Friend and Neighbor," May 19, 1960, DCFB Archive.

30. Patriotism was clearly an important motivation. Council members liked to point out that the American agriculture sector, comprising roughly 8 percent of the nation's population, was producing sufficient products to feed the nation and export overseas. In contrast, they noted that in communist nations, 45 percent of the population produced barely enough for their own domestic needs.

31. U.S. Farm Policy Council, "Statement of Functions and Objectives."

32. U.S. Farm Policy Council, "Proposed Public Relations Program for Agriculture" and "Statement of Functions and Objectives"; "Report of Farm Policy Council Meeting at Des Moines, Iowa, December 8 & 9, 1960"; letter, Ed O'Shea to Rasmusen, January 30, 1960; letter, Ralph Cole to "Fellow Farmers and Friends of Agriculture," December 23, 1960; "Report of Farm Policy Conference, Holdrege, Nebraska, March 4, 1960"; Resolutions by Farm Policy Conference, Holdrege, Nebraska, March 4, 1961; "Prairie Farmer's Farm Outlook Letter," address by Paul C. Johnson to Federal Land Bank Conference, Louisville, Kentucky, August 30, 1961, DCFB Archive. *DeKalb Daily Chronicle*, July 19 and 20, 1960, One of the most sympathetic articles to appear at the time was Mary Conger, "The Farmer's Side of the Case," *Saturday Evening Post* April 9, 1960. Conger was also a featured speaker at the May 26 meeting.

33. Author's interview with Howard Mullins, July 9, 2003.

34. Minutes, board of directors meeting of the DeKalb County Farm Bureau, July 25, 1960, August 22, 1960; U.S. Farm Policy Council, *Newsletter*, August 9, 1961; letter, Ralph H. Cole to Russel Rasmusen, November 8, 1961; letter, Russel Rasmusen to Ralph Cole, December 13, 1961, DCFB Archive.

35. Mullins's hasty threat seemed to many outside DeKalb County as though he was flaunting his farm bureau's success. Yet the DeKalb County Farm Bureau was the most powerful in Illinois and was emerging as one of the most influential farm bureaus in the nation by the early 1960s. Growing resentment of that success, coupled with Mullins's comment and the Bureau's national profile in agricultural politics, triggered a quarter century of hard feelings between the DeKalb County Farm Bureau and a generation of state and national farm bureau leaders that made it difficult for the DeKalb County Farm Bureau or its representatives to sustain their leadership roles outside DeKalb County. Author's interview with Doug Dashner, January 2005.

36. Author's interview with Allan Aves, April 24, 2003; interview with Jeff and Mary Lu Strack, July 22, 2003.

37. The Feed-Grain Bill, H.R. 4510, and its counterpart in the Senate, passed their respective houses and went to conference committee in March 1961.

38. Letter, Ralph H. Cole to "Fellow Farmers and Friends of Agriculture," March 16, 1961, DCFB Archive.

39. Cochrane, a University of Minnesota professor, was an ardent council supporter and a frequent speaker at council events.

40. Letter, Robert K. Buck and Howard Mullins to President John F. Kennedy, June 7, 1962; letter, Russel Rasmusen to Robert K. Buck, June 1, 1962, DCFB Archive. A survey of 57,000 farmers in twenty-five states sponsored by the DeKalb County Agricultural Association revealed that 72 percent of respondents favored a voluntary feed-grain reduction program.

41. E. E. Golden interview of Howard Mullins, August 23, 1979, DCFB Archive. Author's interview with Howard Mullins. U.S. Farm Policy Council, "Goals and Objectives Meeting, DeKalb, Illinois, August 31, 1961," DCFB Archive.

42. *DeKalb Daily Chronicle*, April 27 and June 15, 1962. *Prairie Farmer*, June 2, 1962. DeKalb County Farm Bureau Golden Anniversary Celebration Program, "Progress Through Fifty Years of Farm Bureau, 1912–1962"; *DeKalb County Farmer*, April–August, 1962; letter, Howard Mullins to Al Golden, June 19, 1962; "Proposed Line-up for Parade for 50th Anniversary of the DeKalb Farm Bureau," June 14, 1962, DCFB Archive. Mrs. Lloyd B. Waldee won the theme contest.

43. *DeKalb County Farmer* 50:2 (February 1962).

44. See, generally, "Thoughts . . . On Farm Safety," a regular feature of the DeKalb Farm Bureau's monthly publication, *DeKalb County Farmer*, during the 1960s; *DeKalb County Farmer* 38:10 (October 1951), 49:7 (July 1962); author's interview with Howard Mullins.

45. *DeKalb County Farmer* 56:2 (February 1962). Letter, J. E. Hodges to Robert B. Rogers, July 15, 1965; letter, Lawrence Eaton to Henry C. Garlieb, June 7, 1967, DCFB Archive. Locker Service assets were distributed to shareholders. Despite suffering losses for nearly four years, in the final accounting the service was only $17.61 in debt after paying the costs of liquidation. Donald Upston, administrator, "Administrator's Notice of Final Assessment, July 20, 1965," DCFB Archive.

46. See, generally, *DeKalb County Farmer*, 1960–1970.

47. *DeKalb Daily Chronicle*, March 2, 1967.

48. See, generally, *DeKalb County Farmer*, 1960–1970.

49. Criticism of the American Farm Bureau Federation had a long history. See, for example, Wesley McCune, *The Farm Bloc* (1943; reprint, New York: Greenwood Press, 1968); Grant McConnell, *The Decline of Agrarian Democracy* (Berkeley: University of California Press, 1953; reprint, 1959 and 1969); Samuel R. Berger, *Dollar Harvest: The Story of the Farm Bureau* (Lexington, Mass.: D.C. Heath, 1971). In 2000, the investigative television news magazine *60 Minutes* ran a story highly critical of the American Farm Bureau Federation and alleging, among other things, that the it was closely

allied with agribusiness interests; its leadership paid little attention to improving the condition of poor farmers; and that it was a propaganda machine for a variety of extreme Right-Wing viewpoints.

50. *DeKalb County Farmer* 55:9 (September 1968); *DeKalb Daily Chronicle,* September 20, 1968.

51. Thomas H. Roberts, Jr., "World Food Supply-Demand Prospects," paper delivered to the Seventh International Commodities Conference, Chicago, December 12, 1977, DCFB Archive.

52. Roe C. Black, "The New Era for Agriculture," *Farm Journal* (March 1977); Gilbert C. Fite, *American Farmers: The New Minority* (Bloomington: Indiana University Press, 1981); Bryan Jones, *The Farming Game* (Lincoln: University of Nebraska Press, 1983); E. E. Golden, "European Trade and Marketing Trip," November 12, 1963; Roberts, "World Food Supply-Demand Prospects," DCFB Archive.

53. "Old MacDonald Sold His Farm," *Economist* (London), December 1, 1984, pp. 31–39.

54. E. E. Golden, "European Trade and Marketing Trip," November 12, 1963, DCFB Archive. Author's interview with E. E. Golden.

55. The DCX board of directors consisted of Bureau president Howard Mullins as DCX president; Donald Chaplin, who succeeded Mullins as Bureau president, as DCX vice president; longtime Bureau director George Tindall as DCX "second vice president"; Paul Montavon, one of the key figures in the Farm Policy Council movement, as DCX secretary; and another Bureau director, Raymond Willrett, as DCX treasurer; Allan Aves, who succeeded Chaplin as Bureau president, Ralph Montgomery, Robert Diedrich, and Richard Walters completed the board. "DeKalb County Exports, Inc., Unanimous Written Consent to Action of Board of Directors in Lieu of Meeting" (1974). DCX changed its address in 1974 to that of its new facility in Ottawa, Illinois. Memorandum, to DeKalb County Farm Bureau from DeKalb County Exports, "Subject: New Address" (1974). DCX originally issued 11,000 shares with a par value of $100. DeKalb County Exports, "Report of Issuance of Shares" (1974), DCFB Archive.

The Bureau directors were interested in financing the proposed DCX expansion and asked the Bureau's attorneys for an opinion about whether they could pledge Rural Improvement Society assets as security for loans to DCX. "We have told you that it is proper," the attorneys replied, "for Rural Improvement Society to use its assets for the benefit of the Farm Bureau, since the Farm Bureau is another [tax] exempt organization. The Farm Bureau, in turn, is free to invest its funds in DeKalb Co. Exports, since, in addition to being just like an investment in any other business enterprise, the investment is in furtherance of the Farm Bureau's exempt purposes. [W]e advise you that the investment in [DCX] should be made by the Farm Bureau, and that the money needed by the Farm Bureau for this purpose should be borrowed by the Farm Bureau." Letter, Loren E. Juhl to Melvin Hayenga, May 16, 1973, DCFB Archive.

56. On June 25, 1973, the DeKalb County Farm Bureau agreed to guarantee a loan for operating capital for any amount up to $1 million. "Resolution of DeKalb County Farm Bureau Board of Directors," June 25, 1973, DCFB Archive.

57. *DeKalb County Farmer* 1:2 (June 1973) and 1:4 (November 1973). In 1973, the *DeKalb County Farmer* changed format from a monthly to a quarterly journal. When the format changed, the *Farmer* also changed the numbering of its volumes—hence the first quarterly that appeared in 1973 began with volume 1. The DCX sales team also marketed their products by sending samples, about two pounds each, to potential buyers. See for example, letter, Sam Huey to Webster International Corporation, August 12, 1974, DCFB Archive.

58. *DeKalb County Farmer* 1:2 (June 1973) and 1:4 (November 1973). DeKalb County Exports, "Trial Balance," March 31, 1973. A typical contract is illustrated by the one between DCX and P.P.A. Incorporated, of Van Wert, Ohio. Under the contract terms, DCX agreed to buy 50,000 bushels of Waxy Maize Corn with 15.5 percent moisture and furnish the barge for the pickup. Contract, DeKalb County Exports with P.P.A., Incorporated, April 9, 1974. Memorandum, Sam Huey to Sales Committee, Howard Mullins, and James Duvall, July 18, 1974, DCFB Archive.

59. *DeKalb County Farmer* 1:3 (Summer/August 1973), 1:4 (Fall/November 1973), 2:1 (Spring/June 1974). The U.S. Department of the Interior was concerned about the environmental impacts of the proposed facility. The first load of construction steel arrived on the site the day the interior department withdrew its objections. Memorandum, Howard Mullins to export directors, September 28, 1973, DCFB Archive.

60. *DeKalb County Farmer* 2:2 (Spring/June 1974). Letter, Loren E. Juhl to Melvin Hayenga, August 22, 1974; contract, "Agreement; DeKalb County Exports, and Sam Huey Company," February 14, 1972. DCFB Archive.

61. Memorandum, DeKalb Country Exports to board of directors of DeKalb County Farm Bureau, "October Report," October 29, 1974; letter, DeKalb County Exports to Farm Bureau board, April 28, 1975, DCFB Archive.

62. Memorandum, DeKalb County Exports to board of directors, DeKalb County Farm Bureau, "March Report," March 21, 1975. The dryer and related storage tanks cost nearly $1 million. DeKalb County Farm Bureau, "Resolution," August 25, 1976 DCFB Archive.

63. In 1976, the DCX board included longtime Farm Bureau leaders Joseph Faivre, Carl Hill, and Donald Huftalin. Aves and Willrett stayed on. DeKalb County Exports, "Written Consent of Shareholder to Corporate Action in lieu of Annual Meeting," January 26, 1976, DCFB Archive.

64. Author's interview with Allan Aves.

65. DCX elevator volume increased from 10,280,000 bushels in 1979 to 23,348,000 bushels in 1980. Likewise, sales increased from $38 million in 1979 to more than $100 million the following year. DCX paid $380,000 in premiums to area farmers for specialty grains in 1980. DCX even opened a sales office in Tokyo, Japan, under the management of newly appointed vice president of DCX Japan, Jack Yamashita. DeKalb County Farm Bureau, "Our 1980 Family Report," (1981), DCFB Archive.

66. The collapse and sale of DCX seriously weakened the Bureau's finances. At the beginning of the DCX venture, the Bureau and the Rural Improvement Society had about $11 million in total assets, most of it consisting of stock in the DeKalb County Agricultural Association and the Association's subsidiary companies. After guaranteeing the DCX loans and assuming its operational debt, the Bureau and the Society had about $5 million left. The final break between the Bureau and DCX came when the DCX board of directors asked the Bureau to pledge the remaining $5 million to guarantee a huge sale of grain to Syria. Such a move would have put all the remaining Bureau assets at risk, something the Bureau board of directors refused to do, having already lost around 50 percent of their assets. Although it was popular at the time, and even currently, to suggest that DCX was a gamble that nearly "bankrupted" the Bureau, the Bureau came away from the business with assets of about $10 million, an amount that was still a far greater endowment than many other farm bureaus in Illinois, or even nationally, possessed. Author's interview with Doug Dashner, January 2005.

67. Author's interview with Allan Aves.

68. The history of DCX was compiled from the following interviews: Mullins, 1979 (with Golden) and 2003 (with author); Allan Aves, 1983 (with Golden) and 2003 (with author); Raymond Willrett, 1983 (with Golden); author's interview with George

Tindall; author's interview with Ken Barshinger, May 7, 2003; author's interview with Golden, 2003. Another valuable source of general information is *DeKalb County Farmer*, for the period from 1971 through 1982, and DeKalb County Farm Bureau annual reports, for the same period.

69. Press release, "History of Country Pride Meats" (n.d.), DCFB Archive. Members of the board of directors were Roger Steimel, president; John Huftalin, vice president; Paul Schweitzer, secretary; Kenneth Rehn, treasurer; and Roger Hueber, Norman Johnson, and David Wirsing, directors. All of the board members were livestock producers.

70. Halsey was later hired by DCMS as the manager of Country Pride Meats.

71. *DeKalb Chronicle*, November 18, 1980.

72. DCMS and Farm Bureau officials insisted that their new facility was not a "slaughterhouse" but rather a "processing facility."

73. Country Pride Meats opened on October 14, 1980, with a grand public opening on November 22, 1980. *DeKalb Daily Chronicle*, November 18, 1980. Press release, "History of Country Pride Meats" (n.d.), DCFB Archive. The address for the new facility was 108 Harvestore Drive, DeKalb. The marketing service sold its Country Pride meat at that location and at its retail store at 310 North Sixth Street, DeKalb, next door to the Farm Bureau headquarters.

74. Allan Aves, "Opening Statement" (November 22, 1981), DCFB Archive.

75. *DeKalb Daily Chronicle*, November 18, 1980.

76. Shortly after DCMS began to market "Country Pride" meats, it discovered that the name "Country Pride" was already a trademark for another company. Consequently, DCMS changed its brand of products to "Country Style."

77. "DeKalb's Country Style Meats" (n.d.); "History of DeKalb's Country Style Meats" (n.d.); Mariam Nelson, press release, "Country Pride Meats" (n.d.); DeKalb County Farm Bureau, "Our 1980 Family Report" (1981), DCFB Archive. *DeKalb Daily Chronicle*, November 18, 1980.

78. Author's interview with Howard Mullins.

6—The Future

1. The official address is 1350 West Prairie Drive, Sycamore, Illinois. The first office was located in a storefront on North Third Street in DeKalb and operated from 1912 to 1917; the second was the old schoolhouse on North Fifth Street, in DeKalb, that housed the Bureau from 1917 to 1951; and the third office was on North Sixth Street, DeKalb, which the Bureau occupied from 1951 to 1996.

2. *Point of View* 17:13 (October 1, 1996). *Point of View* is the DeKalb County Farm Bureau's publication that is the latest incarnation of the *DeKalb County Farmer*, the Bureau's monthly newsletter that Eckhardt originally created in 1912. Souvenir Booklet, dedication ceremony of the DeKalb County Farm Bureau Building, August 29, 1951, DCFB Archive.

3. The land immediately north and east of the building is zoned for a light industrial park, while the fields to the west and south are likely to become commercial or residential property as well.

4. The building has eight separate office suites. In addition to the Farm Bureau, the facility houses Northern FS (the petroleum and farm supply cooperative), DeKalb County Farm Business Management, agents of Country Companies Insurance, the DeKalb County Soil and Water Conservation District and the DeKalb County Natural Resources Conservation Service, the DeKalb County Farm Service Agency (a branch of the U.S. Department of Agriculture charged with managing federal loans

and payments to farmers), Farm Credit Services of Northern Illinois (which arranges financing for farmers), the University of Illinois Extension Service, and the Farm Service Department of the National Bank and Trust Company.

5. See, generally, John Nesbit, *Megatrends: Ten New Directions Transforming Our Lives* (New York: Warner, 1982).

6. See, generally, Don A. Dillman, "The Social Environment of Agriculture in Rural Areas," in R. J. Hildreth, Kathryn L. Lipton, Kenneth C. Clayton, and Carl C. O'Connor, eds., *Agriculture and Rural Areas Approaching the Twenty-First Century* (Ames: Iowa State University Press, 1988).

7. Battelle Memorial Institute, *Agriculture 2000: A Look at the Future* (Columbus, Ohio: Battelle Press, 1983), p. 82.

8. See, for example, Indur M. Goklany, "Applying the Precautionary Principle in a Broader Context," in Julian Morris, ed., *Rethinking Risk and the Precautionary Principle* (Oxford: Butterworth-Heinemann, 2000), and *The Precautionary Principle: A Critical Appraisal of Environmental Risk Assessment* (Washington, D.C.: Cato Institute, 2000); Drew L. Kershen, "The Risks of Going Non-GMO," *Oklahoma Law Review* 53:4 (2000): 631–52; U.S. Department of Agriculture, *Food and Agricultural Policy: Taking Stock for the New Century* (Washington, D.C.: Government Printing Office, 2001); Janet Carpenter and Leonard P. Gianessi, *Agricultural Biotechnology: Updated Benefit Estimates* (Washington, D.C.: National Center for Food and Agricultural Policy, 2001).

9. See, for example, Roger E. Meiners and Bruce Yandle, eds., *Agricultural Policy and the Environment* (Lanham, Md.: Rowman & Littlefield, 2003).

10. U.S. Department of Agriculture, *Food and Agriculture Policy*, p. 75.

11. Gary D. Libecap, "Agricultural Programs with Dubious Environmental Benefits: The Political Economy of Ethanol," in Roger E. Meiners and Bruce Yandle, eds., *Agricultural Policy and the Environment* (Lanham, Md.: Rowman & Littlefield, 2003).

12. Don Paarlberg, *American Farm Policy: A Case Study of Centralized Decision-Making* (New York: John Wiley and Sons, 1964), p. 3.

13. Philip S. Foner, ed. *Basic Writings of Thomas Jefferson* (New York: Wiley, 1944), pp. 161–62.

BIBLIOGRAPHY

Published Books and Articles

Bakken, Henry H., and Marvin A. Schaars. *The Economics of Cooperative Marketing*. New York: McGraw-Hill, 1937.

Barger, Harold, and Hans L. Landsberg. *American Agriculture, 1899–1939*. New York: National Bureau of Economic Research, 1942.

Battelle Memorial Institute. *Agriculture 2000: A Look at the Future*. Study by Columbus Division, Battelle Memorial Institute. Columbus, Ohio: Battelle Press, 1983.

Bay, Edwin. *The History of the National Association of County Agricultural Agents, 1915–1960*. Springfield, Ill.: Frye Printing, 1961.

Benedict, Murray R. *Farm Policies of the United States, 1790–1950*. New York: Octagon Books, 1966.

Berger, Samuel R. *Dollar Harvest: The Story of the Farm Bureau*. Lexington, Mass.: D. C. Heath, 1971.

The Biographical Record of DeKalb County, Illinois. Chicago: S. J. Clarke, 1898.

Birkbeck, Morris. *Letters from Illinois*. New York: Da Capo Press, 1970.

Black, Roe C. "The New Era for Agriculture." *Farm Journal* (March 1977).

Boies, Henry Lansom. *History of DeKalb County, Illinois*. Chicago: O. P. Bassett, 1868. Reprint, Evansville, Ind.: Unigraphic, 1973.

Bowers, William L. *The Country Life Movement in America, 1900–1920*. Port Washington, N.Y.: Kennikat Press, 1974.

Boyle, James E., Ron G. Klein, and Stephen Bigolin. "DeKalb County Courthouse." Historical brochure (n.d.).

Burritt, Maurice C. *The County Agent and the Farm Bureau*. New York: Harcourt, Brace, 1922.

Carney, Mabel. *Country Life and the Country School: A Study of the Agencies of Rural Progress and of the Social Relationship of the School and the Community*. Chicago: Row, Peterson and Company, 1912.

Carpenter, Janet, and Leonard P. Gianessi. *Agricultural Biotechnology: Updated Benefit Estimates*. Washington, D.C.: National Center for Food and Agricultural Policy, 2001.

Cochrane, Willard W. *The Development of American Agriculture: A Historical Analysis*. Minneapolis: University of Minnesota Press, 1993.

Cochrane, Willard W., and Mary E. Ryan. *American Farm Policy, 1948–1973*. Minneapolis: University of Minnesota Press, 1976.

Conger, Mary. "The Farmer's Side of the Case." *Saturday Evening Post* (April 9, 1960).

"County Estimates for Twenty Years Ending January 1, 1945." *Illinois Agricultural Statistics*. Springfield: Illinois Cooperative Crop Reporting Service, 1951.

Crabb, A. Richard. *Hybrid Corn Growers: Prophets of Plenty*. New Brunswick, N.J.: Rutgers University Press, 1948.

Cunningham, J. B., E. N. Searls, and O. B. Brown. *Farm Business Report, 1944: Farming-Type Area Two, Northwestern Mixed Livestock Area*. Urbana: University of Illinois College of Agriculture, Extension Service, 1944.

Cunningham, J. B., E. L. Sauer, and E. N. Searls. *Farm Business Report, 1943: Farming-Type Area Two, Northwestern Mixed Livestock Area.* Urbana: University of Illinois College of Agriculture, Extension Service, 1943.

Danbom, David B. *Born in the Country: A History of Rural America.* Baltimore: Johns Hopkins University Press, 1995.

———. *The Resisted Revolution: Urban America and the Industrialization of Agriculture, 1900–1930.* Ames: Iowa State University Press, 1979.

Davis, Joseph. "An Evaluation of the Present Economic Position of Agriculture by Regions and in General." *Journal of Farm Economics* 15:2 (April 1933).

Davy, Harriet Wilson. *From Oxen to Jets: A History of DeKalb County, 1835–1963.* Dixon, Ill.: Rogers Printing, 1963.

The DeKalb Chronicle Illustrated Souvenir Edition, 1899. Reprint, Evansville, Ind.: Unigraphic, 1981.

DeKalb County Home Bureau. "Scrapbook, 1939–1940, Other Early Years Included." DeKalb County Extension Office, DeKalb, Ill.

Deuel, Thorne. "The Hopewellian Community." *Hopewellian Communities in Illinois, Scientific Papers.* Vol. 5. Springfield: State of Illinois, 1952.

Dillman, Don A. "The Social Environment of Agriculture and Rural Communities." In R. J. Hildreth, Kathryn L. Lipton, Kenneth C. Clayton, and Carl C. O'Connor, eds., *Agriculture and Rural Areas Approaching the Twenty-First Century.* Ames: Iowa State University Press, 1988.

Dondore, Dorothy Anne. *The Prairie and the Making of Middle America: Four Centuries of Description.* Cedar Rapids, Iowa: Torch Press, 1926.

Dyson, Lowell K. *Farmers' Organizations.* New York: Greenwood Press, 1986.

Eckhardt, William G. "Hiring the County Expert." *Prairie Farmer,* January 1, 1913.

———. "How Some Concerns 'Educate' Farmers." *Prairie Farmer,* January 15, 1913.

———. "Teaching Farmers by the DeKalb County Plan" *Prairie Farmer,* January 1, 1913.

Eubanks, Mary. "An Interdisciplinary Perspective on the Origin of Maize." *Latin American Antiquity* 12:1 (2001).

———. "The Mysterious Origin of Maize." *Economic Botany* 55:4 (2001).

Fite, Gilbert C. *American Farmers: The New Minority.* Bloomington: Indiana University Press, 1981.

Fitzgerald, Deborah K. *The Business of Breeding: Hybrid Corn in Illinois, 1890–1940.* Ithaca, N.Y.: Cornell University Press, 1990.

Foner, Philip S. ed. *Basic Writings of Thomas Jefferson.* New York: Wiley, 1944.

Gerhard, Frederick. *Illinois as It Is.* Chicago: Keen and Lee, 1857.

Glazer, Sidney. "The Rural Community in the Urban Age." *Agricultural History* 23:2 (April 1949).

Goklany, Indur M. "Applying the Precautionary Principle in a Broader Context." In Julian Morris, ed., *Rethinking Risk and the Precautionary Principle.* Oxford: Butterworth-Heinemann, 2000.

———. *The Precautionary Principle: A Critical Appraisal of Environmental Risk Assessment.* Washington, D.C.: Cato Institute, 2000.

Golden, Elroy E. "Data Relating to the Farm Problem." DeKalb, Ill.: Cooperative Extension Service, DeKalb County, 1960.

Grogan, Robert M. "Present State of Knowledge Regarding the Pre-Cambrian Crystalices of Illinois," *Short Papers on Geological Subjects* 157 (1950).

Gross, Lewis M. *Past and Present of DeKalb County, Illinois.* Chicago: Pioneer Publishing, 1907.

Heichel, G. H. *Comparative Efficiency of Energy Use in Crop Production.* Bulletin 739, Connecticut Agricultural Experiment Station, New Haven (November 1973).

Hill, James J. "Highways of Progress; What WE Must Do to be Fed." *World's Book* 19-20 (November 1909–May 1910).

Horton, Donald C., Harald C. Larsen, and Norman J. Wall *Farm Mortgage Credit Facilities in the United States*. U.S. Department of Agriculture Miscellaneous Publication No. 478. Washington, D.C.: Government Printing Office, 1942.

Illinois Archeological Survey. *Illinois Archeology, Bulletin 1*. Urbana: University of Illinois, 1973.

Illinois Department of Agriculture. *Illinois County Agricultural Statistics, DeKalb County Bulletin C-11*. Springfield: Illinois Department of Agriculture, 1965.

Johnson, H. Thomas. *Agricultural Depression in the 1920s: Economic Fact or Statistical Artifact?* New York: Garland, 1985.

Johnson, P. E., J. B. Cunningham, and E. M. Hughes. *Farm Business Report, 1940: Farming-Type Area Two, Northwestern Mixed Livestock Area*. Urbana: University of Illinois College of Agriculture, Extension Service, 1940.

Johnson, P. E., J. B. Cunningham, and W. D. Buddemeier. *Farm Business Report, 1941: Farming-Type Area Two, Northwestern Mixed Livestock Area*. Urbana: University of Illinois College of Agriculture, Extension Service, 1941.

Jones, Bryan. *The Farming Game*. Lincoln: University of Nebraska Press, 1983.

Kellogg, Jesse C., and Henry Lansom Boies. *DeKalb County History, 1871*. Genoa, Ill.: Thompson and Everts, 1871.

Kelsey, Lincoln David, and Cannon Chiles Hearne. *Cooperative Extension Work*. Ithaca, N.Y.: Comstock, 1955.

Kershen, Drew L. "The Risks of Going Non-GMO." *Oklahoma Law Review* 53:4 (2000).

Kile, Orville M. *The Farm Bureau Movement*. New York: Macmillan, 1921.

———. *The Farm Bureau through Three Decades*. Baltimore: Waverly Press, 1948.

Kouba, Leonard J. *Resource Use and Conservation in DeKalb County*. DeKalb, Ill.: DeKalb County Conservation Education Committee Print, 1963.

Lacey, John J. *Farm Bureau in Illinois: History of Illinois Farm Bureau*. Bloomington: Illinois Agricultural Association, 1965.

Lawrence, William S., and Associates, Inc. *Comprehensive Plan for DeKalb County, Illinois*. United States Housing and Urban Development Project No. P-268. DeKalb, Ill.: DeKalb County Planning Commission, 1971.

Laws of the State of Illinois, 1913. Springfield: Illinois State Journal Co., State Printers, 1913.

Libecap, Gary D. "Agricultural Programs with Dubious Environmental Benefits: The Political Economy of Ethanol." In Roger E. Meiners and Bruce Yandle, eds., *Agricultural Policy and the Environment*. Lanham, Md.: Rowman & Littlefield, 2003.

Lindeman, Eduard C. "Social Implications of the Farmers Cooperative Marketing Movement." *Journal of Social Forces* 1:4 (1923).

Mann, Frank I. *Frank Mann's Soil Book*. Chicago: Prairie Farmer, 1912.

Marsh, Charles W. *Recollections, 1837–1910*. Chicago: Farm Implement News Company, 1910.

McConnell, Grant. *The Decline of Agrarian Democracy*. Berkeley: University of California Press, 1953.

McCune, Wesley. *The Farm Bloc*. 1943. Reprint, New York: Greenwood Press, 1968.

Meiners, Roger E., and Bruce Yandle, eds. *Agricultural Policy and the Environment*. Lanham, Md.: Rowman & Littlefield, 2003.

Mendez, Sara Buerer. *Wigwams to Moon Footprints*. Waterman, Ill.: Waterman Press, 1972.

Mosier, J. G., H. W. Stewart, E. E. DeTurk, and H. J. Snider. *DeKalb County Soils: Soil Report No. 23*. Urbana: University of Illinois Agricultural Experiment Station, 1922.

Myers, James E. "Our Prairie Past." *Chicago Tribune Magazine*, August 21, 1977.

Nesbit, John. *Megatrends: Ten New Directions Transforming Our Lives*. New York: Warner, 1982.

"Old MacDonald Sold His Farm." *Economist* (London). December 1, 1984.

O'Neill, William L. *The Progressive Years: America Comes of Age*. New York: Dodd, Mead, 1975.

Paarlberg, Don. *American Farm Policy: A Case Study of Centralized Decision-Making*. New York: John Wiley and Sons, 1964.

Page, Arthur C. *The Illinois Farmer Book of DeKalb County, 1925*. Chicago: Orange Judd Farmer, 1925.

Parker, Harold. "Good Roads Movement." *Annals of the American Academy of Political and Social Science* 40 (March 1912).

Pimentel, David, L. E. Hurd, A. C. Bellotti, I. N. Oka, O. D. Sholes, and R. S. Whitman. "Food Production and the Energy Crisis." *Science* 182 (November 2, 1973).

Portrait and Biographical Album, DeKalb County, Illinois. Chicago: Chapman Brothers, 1885.

Rasmussen, Wayne D. *Taking the University to the People: Seventy-Five Years of Cooperative Extension*. Ames: Iowa State University Press, 1989.

Richardson, John L. "Banker Brown." *Better Crops with Plant Food: The Pocket Book of Agriculture* (January 1928).

Roberts, Thomas H., Jr. *The Story of the DeKalb "AG."* Carpentersville, Ill.: Carlith Printing, 1999.

Sanders, Harry C. "The Legal Base, Scope, Functions, and General Objectives of Extension Work." In Harry C. Sanders, ed., *The Cooperative Extension Service*. Englewood Cliffs, N.J.: Prentice-Hall, 1966.

Schapsmeier, Edward L., and Federick H. Schapsmeier. *Henry A. Wallace of Iowa*. Ames: Iowa State University Press, 1968.

Schlebecker, John T. *Whereby We Thrive: A History of American Farming, 1607–1972*. Ames: Iowa State University Press, 1975.

"Seed for the Victory Corn Crop a Work Done—A Bit of History Made." *Orange Judd Farmer* (July 13, 1918).

Seventeenth Annual Report of the Illinois Farmers Institute: A Handbook of Agriculture. Springfield: Illinois State Journal Company, State Journal, 1912.

Severson, Lewis E. "Some Phases of the History of the Illinois Central Railroad Company since 1870." Ph.D. dissertation, University of Chicago, 1930.

"State Data through 1944" *Illinois Agricultural Statistics*. Springfield: Illinois Cooperative Crop Reporting Service, 1949.

State of Illinois. *Laws of the State of Illinois*. Springfield: H. W. Rokker, 1893.

Strateon, William J., ed. *Blue Book of the State of Illinois, 1929–1930*. Springfield, Ill.: Journal Publishing, 1929.

Strohm, John. "Farm Power: From Muscle to Motor." *Prairie Farmer* 113:1 (January 11, 1941).

Taylor, Henry C., Anne D. Taylor, and Norman J. Wall. *The Story of Agricultural Economics in the United States, 1840–1932*. Ames: Iowa State College Press, 1952.

Terkel, Studs. *Hard Times*. New York: Pantheon Books, 1986.

Thomson, Betty Flanders. *The Shaping of America's Heartland*. Boston: Houghton Mifflin, 1977.

True, Alfred C. *A History of Agricultural Extension Work in the United States from 1785 to 1923*. U.S. Department of Agriculture Miscellaneous Publication No. 15. Washington, D.C.: Government Printing Office, 1928.

True Republican and Sycamore, Illinois, Illustrated Prospectus, 1906. Reprint, DeKalb, Ill.: DeKalb County Historical Society, 1980.

U.S. Congress. *Report of the Country Life Commission*. Senate Document 705, 60th Cong., 2d Sess., 1909.

U.S. Department of Agriculture. *Department of Agriculture Miscellaneous Publication No. 478*. Washington, D.C.: Government Printing Office, 1942.

———. *Food and Agricultural Policy: Taking Stock for the New Century*. Washington, D.C.: Government Printing Office, 2001.

———. *Yearbook of Agriculture, 1935*. Washington, D.C.: Government Printing Office, 1935.

U.S. Department of Commerce. *Population of the United States in 1860; Compiled from Original Returns of the Eighth Census*. Washington, D.C.: Government Printing Office, 1864.

———. *The Seventh Census of the United States, 1850*. Washington, D.C.: Robert Armstrong, 1853.

———. *Statistical Abstract of the United States, 1920*. Washington, D.C.: Government Printing Office, 1921.

U.S. Department of Commerce, Bureau of Census. *Thirteenth Census of the United States, 1910: Abstract of the Census with Supplement for Illinois*. Washington, D.C.: Government Printing Office, 1913.

———. *Fourteenth Census of the United States, 1920: Agriculture*, "Report for the States, The Northern States." Washington, D.C.: Government Printing Office, 1920.

———. *Fifteenth Census of the United States, 1930: Agriculture*, "Report for the States, The Northern States." Washington, D.C.: Government Printing Office, 1930.

———. *Historical Statistics of the United States, Colonial Times to 1970, Bicentennial Edition*. Part 1. Washington, D.C.: Government Printing Office, 1975.

———. *United States Census of Agriculture, 1925*. Washington, D.C.: Government Printing Office, 1925.

U.S. Federal Farm Board. *First Annual Report for the Year Ending June 30, 1930*. Washington, D.C.: Government Printing Office, 1930.

Voters and Tax-Payers of DeKalb County, Illinois. 1876. Reprint, Evansville, Ind.: Unigraphic, 1977.

Wells, George F. "The Rural Church." *Annals of the Academy of Political and Social Science* 40 (March 1912).

Wessel, Thomas, and Marilyn Wessel *4-H, An American Idea, 1900–1980: A History of 4-H*. Chevy Chase, Md.: National 4-H Council, 1982.

Wiebe, Robert H. *The Search for Order*. New York: Hill and Wang, 1967.

Wik, Reynold M. "The Radio in Rural America During the 1920s." *Agricultural History* 55 (October 1981).

Wilcox, Walter W. *The Farmer in the Second World War*. Ames: Iowa State College Press, 1947.

Williams, Robert C. *Fordson, Farmall, and Poppin' Johnny: A History of the Farm Tractor and Its Impact on America*. Urbana: University of Illinois Press, 1987.

Periodicals

Breeder's Gazette
DeKalb County Farmer
DeKalb Daily Chronicle
Economist
Malta Record
Orange Judd Farmer
Point of View
Prairie Farmer
Sycamore True Republican

Archive Collections

Bertha G. Bradt Collection, Regional History Center Archive, Northern Illinois University, DeKalb.
DeKalb County Farm Bureau Archive, Sycamore, Illinois.
Hall Family Collection, Regional History Center Archive, Northern Illinois University, DeKalb.
Illinois Regional Archive Depository, Northern Illinois University, DeKalb.

Interviews

With Author:

Allan Aves
Wayne Bark
Kenneth Barshinger
Joseph Faivre
Elroy E. Golden
Donald Guehler
Keith Jordal
Ronald Kline
Ray Larson
Howard Mullins
Jack Nelson
Jerry Sanderson
Roger Steimel
Jeff and Mary Lu Strack
George Tindall

With Elroy E. Golden:

Allan Aves
Earl Baie
Donald Chaplin
Eva Eckhardt
Melvin Gehlbach
Charles Gunn
Paul Johnson
Guy Lanan
William Lenschow
Donald Mosher
Howard Mullins
Henry W. Parke
Eleanor T. Roberts
Ethel and Edna Schweitzer
Raymond Willrett

INDEX

4-County Service Company, 154
4-H, 128, 135, 162, 175, 177
Afton Center [Farmers'] Alliance, 42
Agricultural Adjustment Act, 151–52
Agricultural Marketing Act, 124
alfalfa, 65, 89
Allen, William, 23
American Association for Highway Improvement, 29
American Farm Bureau Federation, 110, 159–60, 180, 186, 202, 209–10
American Federation of Labor, 36
Anderson, John, 42
Anglo-American Food Committee, 158
anhydrous ammonia, 169
Archer Daniels Midland/Tabor Grain, 222
automobiles, 121
Aves, Allan, 222, 225
Awe, Chris, 60

Baird, Edna Minard, 61
bank crisis, 130–31
Barb City Quartet, 110
Barron, John H., 38
Barshinger, Ken, 230
Bell, Orton, 142
Beloit, Wisconsin, 12
Benefiel, Eva, 91
Benson, Ezra Taft, 181, 185
"Better Farming Association" (Bottineau County, North Dakota), 39
Biester, Charlotte, 138, 151
Binghamton Chamber of Commerce (New York), 38
Birkbeck, Morris, 12
Bissell, Eugene, 219
Black Hawk (Makataimeshekiakiak), 9
Blair, John, 142
Bloomington Moraine, 4
Boies, Henry Lansom, 14
Boone County, Illinois, 118
Bottum, J. Carroll, 199

Boys' Band of Savanna, Illinois, 110
Bradt, Charles, 47, 51, 60
Breeder's Gazette, 76
Broome County, New York, 38
Broome County Farm Bureau, 40
Broughton, Charlie, 23, 55
Brown, Dillon S., 44, 46, 47, 50, 51, 56, 63, 64, 70
Bryant, William, Cullen, 11
Buckingham, J. H., 11

Cahokia, 6
Cape Girardeau County, Missouri, 39
Castle, J. B., 46
Cavelier, Jean, 7
Center, O. D., 49, 50
Chaplin, Donald, 219, 220, 222
Chemung County, New York, 39
Chicago, Illinois, 12
Chicago International Livestock Exposition, 88
Clark, George Rogers, 8
Clinton Township, Illinois, 106
clover, 73–74; seed purchase program, 74–75
Clover Dairy Farm, 24
Cochrane, Willard, 204
Coleman, M. H., 24
Collot, Victor, 10
Columbia University, 30
Committee on Public Information, 98
Connecticut State Board of Agriculture, 30
Cook, Mrs. Willard, 138
Cook, Willard, 150
Coolidge, Calvin, 151
Cooperative Extension Service, 82, 89, 180
corn: domestication of, 6; hybrid, 132–34, 139, 168
Country Life Commission, 25, 233
Country Life Insurance Company, 117

Lightning Source UK Ltd.
Milton Keynes UK
UKHW040745290919
350631UK00003B/175/P